24.88

B.

Create Your Own DVDs

▶▶▶▶▶

▶▶▶▶▶▶▶

▶▶▶▶▶▶

McGraw-Hill/Osborne

New York Chicago San Francisco
Lisbon London Madrid Mexico City
Milan New Delhi San Juan
Seoul Singapore Sydney Toronto

Brian Underdahl

The _McGraw-Hill_ Companies

McGraw-Hill/Osborne
2600 Tenth Street
Berkeley, California 94710
U.S.A.

To arrange bulk purchase discounts for sales promotions, premiums, or fund-raisers, please contact **McGraw-Hill**/Osborne at the above address. For information on translations or book distributors outside the U.S.A., please see the International Contact Information page immediately following the index of this book.

Create Your Own DVDs

1234567890 CUS CUS 019876543
Book P/N 0-07-222732-X   and   CD P/N 0-07-222733-8
parts of
ISBN 0-07-222731-1

Publisher: Brandon A. Nordin
Vice President & Associate Publisher: Scott Rogers
Editorial Director: Roger Stewart
Acquisitions Editor: Megg Morin
Project Editor: Jody McKenzie
Acquisitions Coordinator: Tana Allen
Technical Editor: Jim Kelly
Copy Editor: Carl Wikander
Proofreader: Stefany Otis
Indexer: Irv Hershman
Computer Designers: Tara A. Davis, George Toma Charbak, Elizabeth Jang
Illustrators: Lyssa Wald, Melinda Moore Lytle, Michael Mueller
Series Design: Mickey Galicia, Peter F. Hancik
Cover Design: Tree Hines
Front cover image: Rommel/Masterfile
Back cover image: Elizabeth Knox/Masterfile

This book was composed with Corel VENTURA™ Publisher.

Dedication

To the best builder I've ever known, my good friend Bob Allen.

About the Author...

Brian Underdahl is the best-selling author of more than 65 books. His easy-to-follow writing style makes his subjects capable of being clearly understood by a broad range of appreciative readers. He has been invited as a featured guest on numerous TV shows because of his technical expertise.

Contents

Acknowledgments

I'm a lucky guy. I get to see my name on the cover of a lot of books. But I'm also intelligent enough to know that I'm just a part of the team that makes those books possible. I'd like to thank a whole bunch of people who helped make this book a reality. They include (but aren't limited to) the following:

Brandon A. Nordin, Publisher; Scott Rogers, Vice President & Associate Publisher; Megg Morin, Acquisitions Editor; Tana Allen, Acquisitions Coordinator; and Jim Kelly, Technical Editor.

I'd also like to thank Andy Marken for once again smoothing my way through a bunch of minefields.

In addition, Sonic, Pinnacle, Roxio, Verbatim, Sony, Maxell, Panasonic, and several others provided hardware, software, or supplies that made this project possible. I couldn't have done it without them.

Introduction

I set out to write this book for one simple reason: I want you to have fun making your own movies with your camcorder and your PC. I want to help you get up to speed so that you can easily create movies that your friends and family will actually watch.

Although the title of this book might suggest that it is strictly for people who want to create DVDs, I'm also going to tell you how you can create movies on CDs and videos for your Web site and even make videotape copies for people who still use VCRs. Hey, I know that you might not quite be ready to take the plunge and jump all the way into making actual DVDs yet—that's okay. In this book, you'll learn that you can get your feet wet gradually (probably with the PC you already own). If you decide that making DVDs sounds like something you'd like to do after you've had a little practice, what you'll learn by reading this book will help you make better choices in selecting your hardware and software upgrades.

I've divided this book into three parts. In Part I, I'll give you an introduction to digital video editing, the basics of creating DVDs, and information about how to get your movies into your PC so that you can do something with them. In Part II, I'll show you each of the steps along the way of creating and editing your movies on your PC. I'll make certain that you will gain an understanding of the whole process, so that making movies will seem like the step-by-step logical process that it should be rather than some overwhelming task that is beyond comprehension. Finally, in Part III, I'll show you how to go from the beginning to the end of the movie-making process using several different popular video editing programs. In all likelihood, your current video editor will be one of them, but even if you're still deciding which program to use, you'll find the comparisons among the different programs interesting and useful.

So, please, enjoy yourself! This is going to open up whole new worlds for you.

PART I

00:20:33

Introduction to
DVDs and VCDs

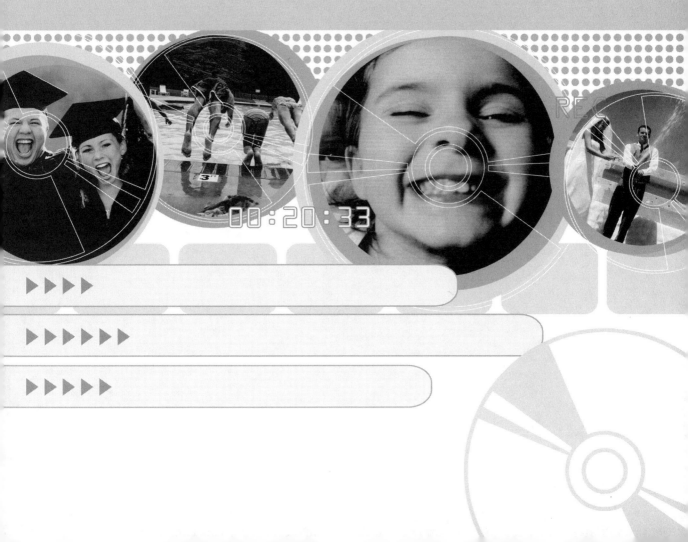

CHAPTER 1 00:20:33

Getting Started
with DVDs and VCDs

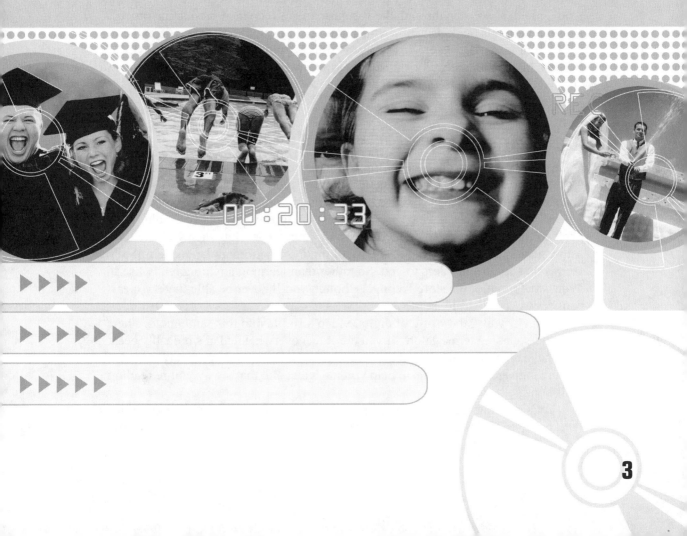

▶▶▶▶

▶▶▶▶▶▶

▶▶▶▶▶

3

I hope you're ready to have some fun, because I intend to make sure that creating your own DVDs and VCDs is an enjoyable experience for you. As you will soon see, these are also very satisfying projects that are made far simpler by the easy-to-use and inexpensive tools that are readily available to you. Today almost anyone can produce top-quality videos with little more than a digital camcorder, a modern PC, the right software, and a basic understanding of how all of these elements come together. It's this last area—managing the hardware and software—where I can do the most to help you, so that's where I'm going to spend most of my time.

Before we go any further I'd like to point out something that I feel is very important— just because I mentioned modern equipment like a digital camcorder, you shouldn't assume that your videos have to be shot on the latest high-tech gear. Actually, that's about as far from my intention as possible. There's no reason why you can't include some of your classic old family photos, home movies, and even audio recordings, too. In fact, as we step through the various parts of the video production process described in this book, I'll show you a number of examples of how you can bring old and new together to create family keepsakes (and other types of videos).

So, let's begin with a look at some of the ways you can use this technology to produce video productions you'll be proud to show.

Understanding DVDs and VCDs

One of the biggest hurdles in learning how to use some of the newest technology is often simply understanding what the "experts" are talking about. I'm sure you've had the same experience as I have had of going into a store to look at something like a new camcorder, seeing several models on display, and really needing some help in figuring out which one will best suit your needs. But when you talk to the store person, it seems like he or she is talking in some foreign language full of odd terms that are meaningless to you. So rather than learning anything really useful, you wander out of the store feeling as though you'll never be able to get a clear answer to your questions.

In many ways, the world of digital video also fits into this same mold. There are odd-sounding terms galore. If you don't feel overwhelmed, it's probably because you have already tuned the babble out. Either way, you could probably use a little plain language explanations to help you navigate, and that's why you're reading this book.

Let's start our plain language explanations by introducing a couple of terms that are used to describe the discs where we'll be storing our videos. These are

DVD and *VCD*—the names of the two most popular types of discs for home video production. Here are some things you'll want to know about these two:

- The term "DVD" is most often identified as standing for *Digital Versatile Disc* or *Digital Video Disc*—there seems to be little real agreement about which definition is correct.

- "VCD" stands for *Video Compact Disc*—essentially a type of CD that holds videos. In fact, these discs are sometimes also called VideoCDs.

- Both DVDs and VCDs are the same physical size as audio CDs.

- To create either a DVD or a VCD, you need the correct type of recorder and recordable discs.

- DVDs generally offer higher quality recordings than do VCDs. This is not an absolute, however, since you can trade off quality for longer recording time if you want.

NOTE *There is a variant of the VCD called S-VCD that has higher quality than standard VCDs (although still not quite up to the level of standard DVDs). Here, too, the tradeoff is shorter recording time on the S-VCD.*

- DVDs are somewhat more expensive to produce than VCDs. That's because DVD recorders are more expensive than CD recorders, and recordable DVD blanks are more expensive than recordable CDs. The gap between the two is narrowing, however.

- Most recordable DVDs hold 4.7 GB of information, while recordable CDs hold 640 to 700MB. This translates to about an hour of video on a typical VCD and up to six hours on a DVD (if you choose a lower quality recording setting).

- DVDs you record will generally play in most set-top DVD players (although, as I'll explain later in this chapter, this capability may depend on the type of disc you use). VCDs will also play in a lot of set-top DVD players, but they're not as widely compatible as DVDs. S-VCDs are even a bit less compatible.

So, now that you have some of the basics under your belt, let's turn to some of the ways you can use these discs.

What You Can Do with DVDs

You're no doubt aware that DVDs have become the preferred format for distributing movies these days. Compared with video tapes, DVDs are more durable, more versatile because of features such as direct access to specific scenes and additional soundtracks in foreign languages, and less susceptible to casual copying. But you may not have thought about some of the different ways you might use recordable DVDs of your own. Well, consider these ideas:

■ You could turn your vacation video into a movie that would actually be interesting enough to keep your viewers from nodding off. Wouldn't it be nice to be able to share the fun and have home movies that people would really enjoy watching? (For example, in Figure 1-1 I'm in the process of editing footage from a recent road trip as I create a video that will have my viewers holding on to the edges of their seats.)

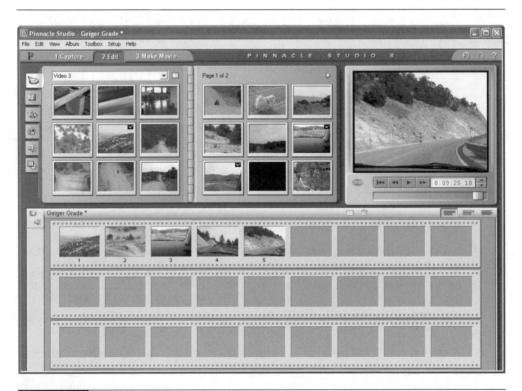

FIGURE 1-1   You can make home videos that people will enjoy watching.

- You could create a recruitment video for your favorite club or organization. In it, you could show some of the fun activities the group engages in, provide some historical background about the group, and provide complete details about how to join.

- You could create a campaign video for a friend or relative who wants to run for local office. This type of video might be especially useful for candidates running a low-budget campaign but still wanting to get their message across to the voters.

- If you've always wanted to start a new career teaching groups of people about a particular interest or hobby of yours, you could produce a video of your "talk" that people could buy. That way you might be able to reach a much larger audience, since you could afford to mail a DVD to places where you probably couldn't get together a group of people large enough to even pay for your travel expenses.

- If you run a business selling antiques, you might want to produce a digital video catalog to show potential customers your wares. Not only could you show all sides of the items as you walked around them, but you could describe the condition of an object using the soundtrack. In addition, your DVD could include easily accessible menus so buyers could quickly navigate to items of interest.

- You could create a promotional video telling about an annual local festival. You might even be able to convince some TV stations in surrounding areas to air parts of your video to help promote the event in their areas.

I could go on, but I think you get the picture. With a little bit of imagination you can probably come up with hundreds of different ways to use DVDs. I'm sure that at least some of the ideas I mentioned made you stop and think, though, didn't they? Well, that's the whole point—DVDs really are pretty versatile!

TIP	*If you make videos you intend to sell, you may want to ask your favorite legal expert for advice about obtaining permission to show people in your videos. It's generally better to deal with these issues ahead of time rather than risk being sued once your video has been released.*

What You Can Do with VCDs

Okay, so the things you can do with recordable DVDs sound pretty good, but what if you're not quite ready to spend a bunch of money for a DVD recorder? Your PC probably already has a CD-RW drive, and maybe you'd like to get your feet wet by starting out with VCDs rather than making the commitment to the more expensive DVD format. Well, that's no problem and indeed may be an excellent way to get started in digital video production.

As I mentioned earlier, VCDs are generally compatible with most newer set-top DVD players. It's true that recordable DVDs can be played in some DVD players that won't play VCDs, but we're talking relatively small numbers here. Unless you're creating videos for commercial purposes—such as videos you intend to sell directly to the public—you may not consider this to be a serious problem.

So, aside from many of the same uses as recordable DVDs, how might you take advantage of the special characteristics of VCDs? Here are some thoughts:

- VCDs can be played in the CD drive on most PCs—even those without a DVD drive. If you're creating videos to share with PC users, VCDs are still likely to be compatible with a broader range of systems than are DVDs.

- Since VCDs are recorded on inexpensive CD-R discs, you can make duplicates of VCDs for far less than the cost of a recordable DVD. In fact, you can probably buy blank CD-R discs for less than the cost of a first-class postage stamp, so handing out copies of your VCD videos at a convention could be a very inexpensive way to promote your business.

- If you're the recruitment chair for a student group, you might create several hundred copies of your club's promotional video on VCDs, then ask the Registrar to include a copy in each new student's information packet. Since most students would have easy access to a PC that could play a VCD (but maybe not a DVD), your high-tech approach to recruiting new members might pay back many times over.

There's no reason why you can't use VCDs as your low-cost entry point into the world of digital video production. Once you know your way around, you'll find that it's a very easy process to step up to making DVDs. And, as you've just seen, VCDs certainly do fill an important niche that even experienced DVD producers may want to consider from time to time.

Now that you have a better understanding of what you can do with DVDs and VCDs, let's move on and take a look at the very important matter of DVD standards.

Understanding DVD Standards

I will have to warn you right now that I'm going to use this section to tell you about a somewhat boring subject—the different standards for recordable DVDs. Frankly, I would just as soon not have to mention this subject, but the reality is that these standards will have a direct impact on your ability to create DVDs that you can share. So please bear with me. I'll keep this short and to the point.

You may recall that there was a standards war that took place when VCRs were first sold to the public. On one side you had Sony with the Betamax, and on the other you had companies like JVC with VHS. The two competing types of VCRs were incompatible with each other, so you had to choose carefully to make sure you were buying (or renting) tapes in the correct format.

One of the interesting facts about competing standards is that the public does not always choose the superior product. In the case of Betamax vs. VHS, most experts at the time thought that Betamax was the better format. The public, however, chose VHS and eventually Betamax VCRs and videotapes disappeared. The lesson here is that popularity is often more important than having the better product.

The important point that applies to our discussion is that there is a similar standards war going on between competing recordable DVD formats. And just to complicate things a bit, there aren't just two formats; rather, there are two main formats—each with two variations—and one minor player you'll definitely want to avoid. So let's take a quick look at these different format standards to help you understand which will be your best choice.

DVD-R/RW

Currently, the most popular recordable DVD format is DVD-R for write-once discs, and DVD-RW for re-writable discs. There are a couple of good reasons for this:

- DVD-R discs are compatible with more set-top DVD players than any other format. This factor all by itself should be very important to you.

- DVD-R blank discs are considerably cheaper than other types of recordable DVDs. (These discs are the ones labeled "for general use.")

- DVD-R/RW drives tend to be less expensive than competing DVD+R/RW drives, since their sales are higher.

When you buy DVD-R discs, be sure to note the quality of the discs you buy. Many of the generic discs can only be written at 1X speed, while the name brand discs (like the Verbatim discs I use) can be written at 2X speed. Look for the 2X label on the disc or jewel case, as shown in Figure 1-2.

DVD+R/RW

Another DVD recordable format is DVD+R for write-once discs, and DVD+RW for re-writable discs. Although it might not seem like much of a difference in nomenclature, the two formats are incompatible, and you must buy the correct discs to match the drive that is installed in your PC.

DVD+R/RW is currently not as popular as DVD-R/RW. Here are some of the reasons why this seems to be so:

- It appears that DVD+R and DVD+RW discs have more compatibility problems with set-top DVD players than do DVD-R/RW discs. It's hard to get accurate statistics in this area, but it may be a concern if you have an older DVD player.

- The first generation of DVD+R/RW drives could not actually create a DVD+R disc that could be read in a standalone DVD player. Instead, you had to use the more expensive DVD+RW discs. The second generation DVD+R/RW drives are said to have corrected this problem.

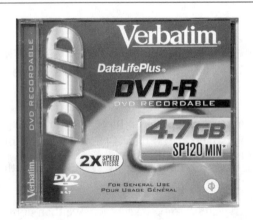

FIGURE 1-2   Be sure to buy high-quality discs, like these from Verbatim, for your digital video projects.

- DVD+R and DVD+RW discs are more expensive than their DVD-R and DVD-RW counterparts.

- DVD+R/RW drives are also more expensive than DVD-R/RW drives. I've been told that a single manufacturer builds all of the DVD+R/RW drives, and everyone else who sells them buys from that one company— thus creating the higher prices.

DVD-RAM

The third recordable DVD format, DVD-RAM, is not really a contender, but it is still important that you know about it. Here's why you don't want a computer with a DVD-RAM drive (or a bare DVD-RAM drive to install in your PC):

- DVD-RAM discs can only be read in DVD-RAM drives on computers.

- DVD-RAM discs are incompatible with most set-top DVD players.

Okay, so it's not a long list, but what more important reasons do you need to stay away from DVD-RAM? Take it from me, there are plenty of people who are willing to sell you things without considering what you really need. Anyone who tries to talk you into a strictly DVD-RAM drive is not doing you any favors and should be avoided!

TIP	*There is a new category of drives, known as* DVD multi, *that supports DVD-R/RW and DVD-RAM discs. These drives offer several important advantages, including both the compatibility of the DVD-R format and the versatility of DVD-RAM (see Appendix A for more information on DVD multi drives and DVD-RAM). DVD multi drives are definitely worth a consideration.*

Well, the discussion of competing recordable DVD formats wasn't that bad after all, now was it? The bottom line is that, if you have not yet selected a recordable DVD drive, I'd suggest that you go with the DVD-R/RW format. If your PC already has a DVD+R/RW drive, it's probably not worth replacing (unless it's one of those first-generation drives like the HP dvd100, which can't create set-top-compatible DVD+R discs). Just be careful to always buy the discs that are compatible with your DVD recorder.

Understanding the Basics of the DVD/VCD Creation Process

Have you ever noticed how much easier things seem once you learn the basics? Until you know how the various elements of a project will fit together, you tend to see the task as one big, incomprehensible, and almost impossible undertaking. How will you ever get your movies from your camcorder onto a DVD that you can play in a set-top player? It just seems overwhelming, doesn't it?

I think the key to making this whole process easier to understand is to break it down into manageable steps. Once you can see how all of the pieces fit together, you'll be able to clearly understand that creating your own DVDs (or VCDs) is really pretty straightforward. Pretty soon you'll have your first DVD movie that you've made for yourself, and things won't seem nearly so intimidating. By the way, from now on I'll just refer to creating "DVDs" unless I'm talking about something that is specific to VCDs. Since the two types of discs are so similar, you'll be following pretty much the same path regardless of which of them you intend to create at the end of the process. Also, for continuity purposes, I'm going to use one excellent and inexpensive piece of video editing software, *Pinnacle Studio 8*, for my general examples. In Part III, I'll show you the specifics of using several popular video editing programs.

> **TIP** *I think it is important for you to remember that it's going to take a little practice before your movies come out exactly the way you would like. Don't get discouraged if the first few you make aren't masterpieces— I'm sure you'll improve rapidly as you spend a little more time getting comfortable with the whole digital video editing process.*

Creating Your Video Content

Not surprisingly, the first step in creating your DVD is actually creating the video content that you want to use for your movie. You have many different options, of course, such as the Sony DCR-TRV27 mini DV camcorder (which I use) shown in Figure 1-3.

I use a *digital* camcorder. Digital camcorders record your video (and audio) as digital information. That is, everything is reduced to a series of ones and zeros, which very accurately store the sounds and images in a format that can be copied or reproduced as many times as you like without deterioration. In most cases, this also means that the end result is a high-quality video that can easily be stored on a very small tape. For example, Figure 1-4 shows that two of the mini DV tapes easily fit within the palm of my hand.

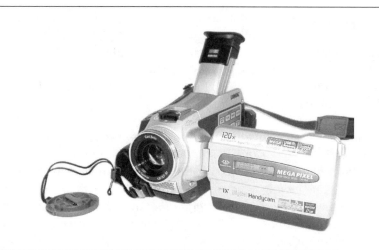

FIGURE 1-3   This Sony digital camcorder is an excellent choice for home video creation.

Mini DV may be the most popular digital camcorder format, but it is not the only one available. Digital 8 uses a slightly larger tape cassette, and micro MV uses an even smaller one. None of this really matters all that much in terms of how you create your video content or the manner in which you will later transfer that content to your PC for editing.

In addition to digital camcorders, there are the older *analog* models, such as the VHS-C and Hi8. Rather than converting your movies to computer data, these

FIGURE 1-4   Mini digital video cassettes are a very compact way to store digital video.

camcorders store the video and audio directly onto the tape. This is the same recording method that is used in VCRs. Analog recordings can suffer a quality loss when the tapes are played many times or when produced by duplication. This characteristic should not be a significant issue if you are transferring a video you've shot from your analog camcorder (or your VCR) to your PC, since once the video is in your PC, it is stored in digital format.

| TIP | *No matter what type of equipment you use to produce your original video, it's good to remember that you can edit the video far more effectively on your PC than you can in the camcorder. Leave some extra footage at the beginning and ending of scenes so that you'll have plenty of leeway for editing on your PC. And don't forget that a well-edited piece is far more enjoyable to watch than one that simply throws in every second of video "just because it's there."* |

The basics of creating your original video footage can be summed up as follows:

- Digital camcorders generally produce superior quality sound and images.

- If you have the proper video capture hardware on your PC, there's no reason why you cannot also use analog sources such as analog camcorders or VCRs.

- Some rare footage is simply too important to waste even if the quality may not be up to modern standards. For example, your grandfather's old home movies probably won't have the same quality as ones you shoot today with your digital camcorder, but that doesn't mean you should simply toss them out!

Capturing Your Video

Once you have your video content assembled, the next step will be to *capture* that video on your PC. This simply means that you will connect the video source to a special *port* on your PC, then copy the images and sounds into your PC for further processing. I'll tell you more about this step in Chapter 3, but for now let's take a quick look at the basics of video capture on your PC.

Capturing Digital Video

If you have a digital camcorder, your PC can connect to it and then control the playback of the video that is contained on the camcorder. As Figure 1-5 shows,

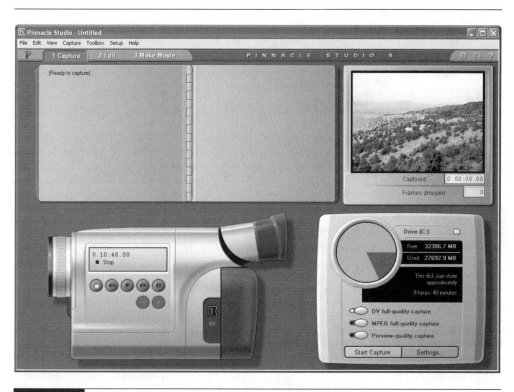

FIGURE 1-5   Capturing video from a digital camcorder is easy with onscreen controls.

the video editing program that you use may even have onscreen controls that enable you to play, pause, rewind, fast-forward, and so on.

In most cases you will probably use an IEEE-1394, FireWire, or i.Link connection between your PC and your digital camcorder. (These are three different names for the same type of connection.) Not all PCs have this type of connection, but it is quite easy to add one if yours does not. An IEEE-1394 connection is really the best method of transferring video between your camcorder and your PC, since it offers very fast transfer rates.

Some digital camcorders also support USB connections to your PC. Unfortunately, USB connections are generally slower than IEEE-1394 connections, so it may take longer to transfer your videos (and you run the risk of introducing errors known as *dropped frames* into the captured video). It's true that the newer USB 2.0 standard is faster than IEEE-1394, but unfortunately it is not yet widely supported by most camcorders—even if your PC is new enough to be equipped with the faster version of USB.

Capturing Analog Video

If you have an analog camcorder or if you want to use recorded video from a VCR, you'll need to use an analog capture method. This is similar to capturing digital video, except that your PC won't be able to control the playback—you'll have to start and stop your camcorder or VCR manually. As Figure 1-6 shows, there isn't a lot of difference in your video editing program (except that there won't be any onscreen playback controls).

You need a different type of connection for capturing analog video signals on your PC. For this, you will need a video capture board that has either *composite video* or *S-Video* inputs, depending on what is available on your video source.

TIP	*If your camcorder or VCR offers an S-Video output, you'll want to use that instead of the composite video output, since the video quality will be superior. In either case, your audio signals will still travel separately from the video signals and will need to be connected to the audio inputs on your video capture board.*

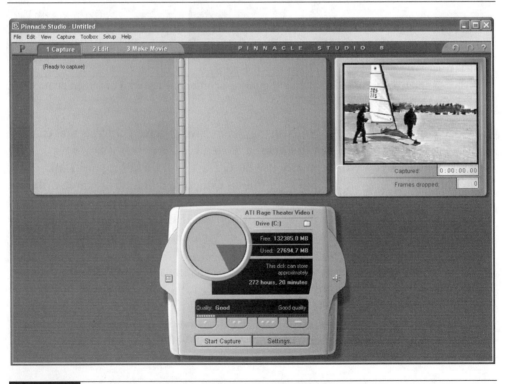

FIGURE 1-6   Capturing video from an analog source is also pretty easy.

It is also possible to capture video through a TV tuner card in your PC. This may be your only option if your VCR lacks composite video or S-Video outputs.

Transferring Old Movie Film to Digital Video

I imagine that some of you may have old movie films (such as 8 MM home movies) that you would like to convert to modern digital video format. The good news is that this is quite possible to do, but the bad news is that you may need to experiment some to get the results you want. Basically, the problem comes down to two issues: first, you need a physical means of doing the transfer; second, you may have some quality issues that will need some tinkering to resolve.

The easiest way to transfer the old movies is to project them onto a screen and then film them with your camcorder (which should be mounted on a tripod to keep it as still as possible). You might also be able to locate a specialized video transfer unit, such as the Raynox RV-1500, which will probably produce superior results. Unfortunately, you may have a difficult time locating a video transfer unit; it is really just a box with a mirror and a frosted glass screen. You set up the film projector to project the images into the box, and the mirror flips the image so it is projected onto the back of the frosted glass screen. Then you focus your camcorder on the screen, start the projector and the camcorder, and wait while the movie is recorded on your camcorder.

The one problem you may encounter using this method is flickering in the video recorded on your camcorder. This flickering results from the frame rate of the old movie being different from the frame rate of your camcorder. The only solution I know about for this problem is to slightly adjust the playback speed on the film projector to see if you can eliminate (or at least reduce) the flickering on your recording.

It may take some extra work, but it is possible to save your old home movies and create those family heirlooms.

Editing Your Movie

Once you've captured all of your video, the next step is to edit it. This is the process of choosing which parts of the video actually belong in your movie. However, to leave the definition there would really only brush the bare surface of the possibilities that you have available for taking your video and making it into a movie that someone will want to watch. In fact, editing is the most fun part of the whole process, because it enables you to really get creative and turn a bunch of boring footage into something that is really special.

Editing your movie can include quite a few different steps. You have to do only one of them—choose the *video clips* that you'll include—but there is a whole world of additional possibilities for you at this point. For example, you can do the following:

- Rearrange the clips so that they appear in a different order than they were shot in. Often this is crucial to telling the story, since you may not have been able to film your shots in the order that best illustrates the story you want to tell.

- You can trim clips to remove extraneous footage at the beginning or the end of a scene. In fact, it's often a very good idea to shoot a few extra seconds both before and after a scene just so you'll be sure to capture everything you really want to include. Then you'll be able to trim properly for a more professional looking final result.

- You can change the speed at which certain clips are played back. For example, you might speed up the pace in a scene shot on the road to increase the excitement, or you might want to slow down a scenery fade out shot at the end of your movie.

- You can add or modify the soundtrack. For example, you can add some music to accompany certain scenes, or you might want to add a *voice-over* narration to tell part of the story.

- You can add titles at the beginning, the ending, or anywhere in between to make your movie appear more professional.

- If you like, you can add *transition effects* between scenes. These are the fancy fades, wipes, dissolves, and so on that you've probably seen in all sorts of movies and TV shows. Transitions can be especially effective when you're moving from one part of your storyline to another, since they give the viewer a visual clue that the next scene is not directly related to the one it follows.

- You can add menus so that viewers can easily navigate through different portions of your movie.

So you see what I mean—editing your movie is where the real fun begins. Just think, with only your camcorder, your PC, and some handy digital video editing software, you'll be able to produce movies that would have cost thousands of dollars and taken the efforts of dozens of people just a few years ago. (Well, maybe the dozens of people is a bit of an exaggeration, but it really is pretty amazing what you can do with this stuff, isn't it?)

Figure 1-7 shows a simple movie I'm editing. To assemble my movie, I begin by dragging the clips I want from the album pages in the upper-left area of the

FIGURE 1-7   Here I've begun the editing process.

window onto the *storyboard*—the filmstrip—near the bottom of the window. I can play any of the clips on my movie by selecting the item I want to play (by clicking it) and using the viewing window at the upper right.

As you learn more about the whole video editing process, you'll find that there are many different ways to work. Some people like to start out with a shooting schedule, and they have the entire movie pretty much planned before they even pick up their camcorder. That's probably a great approach if you're setting out to produce a video that has a specific purpose, such as an advertisement or a recruitment video. But it's also important to remember that you can create videos just for fun, and that sometimes putting too much emphasis on trying to produce a professional-looking movie can just spoil an otherwise fun occasion. For example, if you're taking movies of your family vacation, don't worry about getting every shot just right. Nothing spoils the fun more than having someone tell you to redo something you've just done because they didn't get it framed quite right the first time. Remember that you can always have a lot of fun in the editing phase, and some bloopers will probably make for a more interesting vacation movie, anyway!

Creating Your Disc

Once you have finished your editing, you're almost ready to create your disc. It's a good idea to run through the assembled clips a few times to make certain that everything flows the way you want before you burn your masterpiece to disc—then you can go ahead and burn a sample.

Figure 1-8 shows that I'm about to burn a copy of my ten-minute road trip movie to a VCD. I'm giving the movie one last run through before I click the Create Disc button.

> **NOTE**   *As you may have noticed in Figure 1-8, creating a DVD or VCD copy of my movie isn't my only option. In fact, Pinnacle Studio 8 (and most other video editing programs) gives me a number of different output options. For example, if you want to send Grandma a copy of your movie and you know that she has a VCR but not a DVD player, you could choose to output the movie to tape. Different types of output media have different sets of features, of course, so the videotape you send her won't have the menus you added for your DVD. Still, it's good to know that you have enough options to suit pretty much any need.*

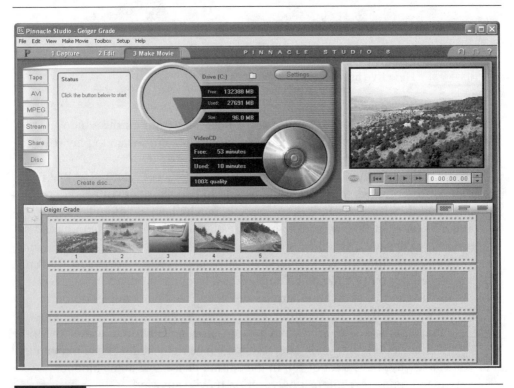

FIGURE 1-8   Now I'm ready to create a disc-based copy of my movie.

Once you begin the disc creation process, it's usually best to just let your computer go at it without additional disturbances. The reason for this is that discs need to be created in a single pass—otherwise they likely won't play in set-top DVD players (and may not even play on your computer). If you have a very fast computer with plenty of memory and a lot of disk space, it's not so likely that anything will disturb the process. However, you'll find that it's probably best to be cautious until you get a good feel for how well your PC handles the whole digital video creation process.

Hardware Basics

Now that you've had a brief introduction to the basics of digital video production, let's move on and take a look at just what hardware you'll need. After all, digital video editing is quite a bit different from things like surfing the Internet, exchanging e-mail, playing solitaire, and balancing your checkbook. I want to make certain that you have what you need to get satisfactory results and enjoy the process.

What Kind of Computer Do You Need?

Digital video editing on your PC is one of those things that has really become practical only during the past few years. Before that, home computers simply weren't powerful enough to do the job—or at least to do it very well. The reason for this is a matter of numbers. Digital video is data, and it is typically huge amounts of data. In order to handle all of this data, your PC needs a fast processor, lots of memory, and plenty of disk space.

Just what does a "fast processor, lots of memory, and plenty of disk space" really mean? Well, you really shouldn't try to do any digital video editing unless your PC meets the following minimum specifications:

- 500 MHz processor
- 128MB of RAM (memory)
- 8 GB of free disk space (will hold a bit over 30 minutes of finished video)

Realistically, I'd suggest doubling all of those numbers at the very least. But even if you do have to upgrade your PC, you should find that it's really not all that expensive to do so. Even the least expensive brand-new PC you can buy today almost certainly has a 1 GHz or faster processor, 128MB of RAM, and 12 GB or more of disk space. Even a minor step up from the basic system should result in a PC that is quite satisfactory.

In case you're wondering, I recently built a new system that handles even the toughest of video editing chores without even breathing hard. It has an AMD

Athlon XP 2200+ processor, 1 GB of RAM, and a 160 GB hard drive. You may not need something that powerful, but it gives you some idea of where you might want to head if you want something a little better than a run-of-the-mill system.

You may have noticed that I haven't mentioned the Mac. That was on purpose. For years it's been almost a religion among Mac users to insist that you had to have a Mac for any type of graphics-related work. Well, when it comes to digital video editing, there's absolutely no reason why you can't use a Windows-based PC just as easily as a Mac. In fact, using a Windows-based PC makes a lot of sense, since they're far less expensive than Macs, have a lot more reasonably priced software available, and are at least as reliable (if not more) than any Mac you can buy. Of course, if you have a Mac and want to use it for digital video editing, that's fine, too. I just don't think that anyone should feel that they have to go out and buy a Mac if they've already got a PC that will work at least as well.

What Computer Add-Ons Do You Need?

In addition to a fast processor, plenty of memory, and lots of free disk space, your PC needs a few extra components that you won't find on most PCs—yet. I expect these will become popular as more people discover the joys of digital video creation on their PCs. But let's take a look at two items you'll definitely need.

Video Capture Options

The first important option you'll need on your PC is one that enables you to transfer the video from your camcorder (or VCR) into your PC. The exact item you'll need depends on the type of camcorder you have:

- Digital camcorders typically use the IEEE-1394 connection I mentioned earlier (this also goes by the names FireWire and i.Link). If your PC is relatively new, it may already have an IEEE-1394 port built in, and you can use that. If it does not have this port, you'll need to add a board inside your computer.

- Analog camcorders and VCRs use either the composite video or the S-Video connections I mentioned earlier. Unless your PC has a TV tuner card, you probably don't have either of these connections and will need to add an adapter board to your PC.

Both IEEE-1394 and video capture adapter boards are pretty easy to find and install. One excellent option you may want to consider is to buy something like *Pinnacle Studio Deluxe*—a complete package that includes the Pinnacle Studio 8 video editing software discussed earlier in this chapter as well as a capture card that gives you both digital and analog inputs and outputs. It has a handy box that sits on top of your PC for easy connections (see Figure 1-9). With this slick package, you can plug in your digital camcorder, your analog camcorder, or even your VCR. And since it has both inputs and outputs, you can even use it to record your movie on that videotape for Grandma. Pinnacle Studio Deluxe is really quite a bargain, too, since everything you need is bundled in one package. You can find more about it at the Pinnacle Web site: www.pinnaclesys.com.

If your PC already has an IEEE-1394 port built in (or you don't need one because you don't have a digital camcorder) and you already have the digital video editing software you need, another solution will enable you to add the composite video and S-Video capture ports without even opening up your PC. The *Pinnacle Bungee DVD* is an external box that connects to your PC through a USB port. As an added bonus, the Pinnacle Bungee DVD even has a built-in TV tuner so you can watch and record TV shows right on your PC's monitor.

 For some reason, it seems as though most digital camcorder manufacturers leave out a very important item when they package and sell you your camcorder. That item is the IEEE-1394 connector cable that you need to transfer the videos from your camcorder to your PC. If you aren't buying a complete video editing package like the Pinnacle Studio Deluxe package, you'll need to buy this cable before you can begin your video editing (the Pinnacle bundle includes the IEEE-1394 cable).

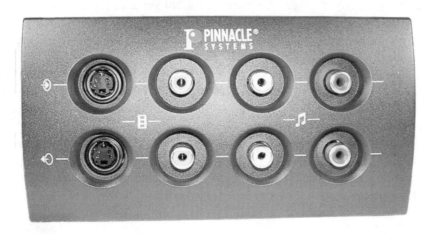

FIGURE 1-9   Pinnacle Studio Deluxe includes a handy I/O panel that sits on top of your PC.

Adding a DVD Burner

The other important item that is still not all that common on most PCs is a DVD burner. Okay, officially this should be called a DVD-R/RW (or DVD+R/RW) drive, but everyone calls these drives "burners" because they use a laser to write the data onto the recordable discs. This is, of course, the drive that you will need if you want to actually create your own DVDs.

> **TIP**   *Remember that you do not need a DVD burner yet if you want to start out your digital video career with VCDs. For that, all you'll need is a standard CD-R/RW drive, which is found on all but the least expensive PCs today.*

As I mentioned earlier, you must choose the recordable DVD format you want to use before you can decide which DVD burner is right for you. If you are just starting out and have decided to go with the DVD-R/RW format, you may want to consider the Pioneer DVR-105 DVD-R/RW drive. This is a brand-new model that replaces the very popular Pioneer DVR-104 model I use.

> **NOTE**   *You may see the Pioneer recordable DVD drives listed as DVR-A05 rather than DVR-105. The "A" designates an OEM drive that is sold in bulk packages rather than in individual retail boxes. The drives are identical, but there may be a difference in both the length of the warranty and in the software bundle (if any) that comes with the drive.*

It really pays to shop around when you're buying a DVD-R/RW drive. The same drive can vary greatly in price—I've seen different vendors price identical drives for anywhere from $250 to $450. In some cases the drive may come packaged with some very minimal software offerings, but this doesn't seem to have too great an affect on the price you pay for the drive.

Well, I hope you've found this brief introduction to the world of digital video editing interesting and informative. I've tried to touch on the important points without getting too technical. Next, we'll have a look at some of your choices in video editing software. I'm going to show you how you can get the job done without breaking the bank, and I'm also going to show you why an inexpensive upgrade may be in order. Finally, I'm going to give you a quick look at a couple of much more expensive options, which you may want to consider once you decide that digital video production is your future path to success and you want to turn pro.

CHAPTER 2

Introduction to DVD Creation Tools

In the last chapter I showed you the basics of the digital video editing and DVD/VCD creation process. One of the important points I discussed was that you need digital video editing software on your PC in order to capture your video, edit that video, and then create your disc. In this chapter, I will tell you more about three different classes of digital video editing software. (I'm not going to show you how to actually create movies in this chapter; that will wait until we reach Part III, after I teach you more about movie making concepts in Part II.)

We will start out by looking at the entry-level editors—such as the one that may have been bundled with your digital camcorder or your DVD-R/RW drive. Next, we will examine a couple of alternatives that represent a significant step up in terms of functionality and flexibility at a very reasonable price. Finally, we will take a brief look at some digital video editing packages that represent a huge jump in both power and cost compared to any of these other options.

Before we get started, I'd like to make a very important point about these different digital video editing and disc creation options. Please keep in mind that you are free to choose whichever editing program you prefer. When comparing the various options that are available to you, think about what you intend to do with your movies. Do you want to use the quickest and most efficient way to get the content onto DVD, do you want to do a lot or little video editing, and do you want to be able to re-edit discs once you've created them? All of these are factors to consider when evaluating the tools you will use.

NOTE *Although I would certainly like to do so, it is simply not possible for me to mention every available digital video editing program in this chapter (or even in this book, for that matter). Rather, I'm going to show you a representative sampling of some of the more popular options. But even though I cannot cover every entry in this area, I am going to show you enough information about the basics of digital video editing so that you can easily get more from the program you choose and end up producing better movies—and that's really what is important, isn't it?*

Okay, so let's begin with a look at your options.

Introducing Some Basic Video Editors

If you are looking for a no-frills approach to digital video editing and disc creation, the basic video editors may be right up your alley. These programs take an approach that could best be described as "hand holding"—something you may or may not like. They are ideal for a beginner to use, and make it quick and easy for you to create DVDs. If you just want to create a movie that takes minimum effort but still looks great, these *template-based* editors may be right for you. (These editors are sometimes called template based because they build your movie based on movie templates, which are essentially predefined movie styles—your content is simply added to an existing movie format.)

Sonic MyDVD

Sonic MyDVD is probably a part of more software bundles than any other digital video editor. This program offers options for customizing your videos, and there's no getting around the fact that its simplicity makes it just about the easiest of all of the video creation tools to use. For example, when you first open the program you have the options shown in Figure 2-1.

NOTE *A new version of Sonic MyDVD with additional features is due to be released shortly after this book is published. As a result, your copy of the program may differ a little from the version you see in the images.*

Sonic MyDVD maintains its simplicity when you click one of the buttons to continue. As Figure 2-2 shows, you still have a few options after your initial choice, and this certainly keeps the whole process of creating your movie much simpler.

TIP *If you are brand new to digital video editing, it's a good idea to run the MyDVD Tutorial, which you'll find in the MyDVD Documentation folder, so that you can take a quick guided tour of Sonic MyDVD. You'll find this folder when you open the MyDVD folder on the Windows Start menu.*

FIGURE 2-1 Sonic MyDVD keeps your options to a minimum.

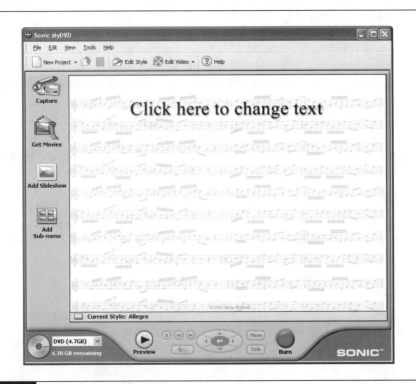

FIGURE 2-2 You just click the option you want to use.

As I mentioned earlier, Sonic MyDVD is a template-based DVD creation application. This will be evident when you have captured some clips, as shown in Figure 2-3. As this figure indicates, Sonic MyDVD's template always builds your movies by tying each separate video sequence to a menu button. These buttons become the mechanism for controlling the playback of your movie. Each button automatically displays the opening shot in the scene to aid in navigation. You can also add descriptive text below each button to further explain what is contained in the scene.

NOTE	*The Sonic MyDVD template allows for up to six buttons on each menu screen. If your movie is split into more than six scenes, additional menu screens will automatically be added to your movie for each set of six scenes that you add.*

To capture your video clips from your digital camcorder or from an analog source (if your PC has an analog video capture board), you click the Capture button to display the Capture window, shown in Figure 2-4. You then use the controls in this window to control your DV camcorder and to select the clips you wish to use.

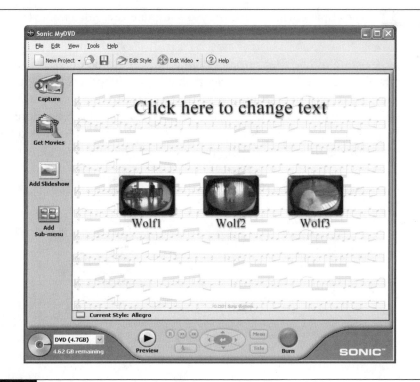

 FIGURE 2-3   Each scene in your movie is automatically attached to a menu button.

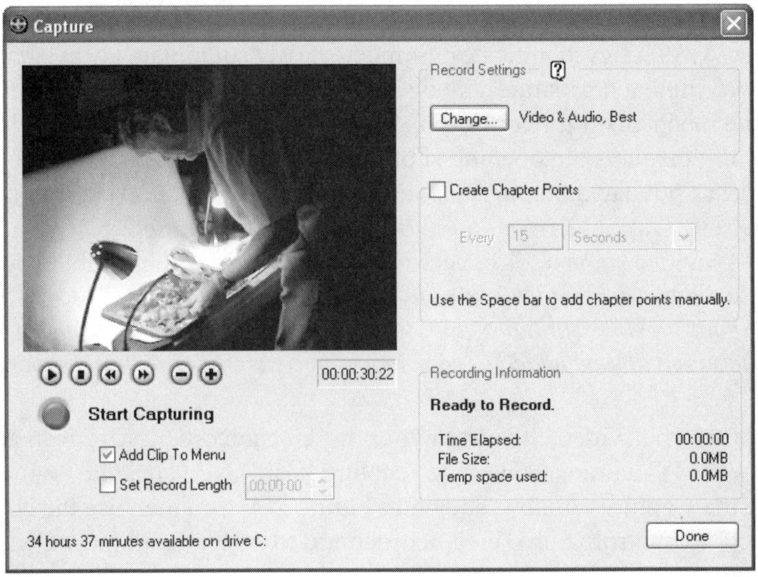

FIGURE 2-4 The Capture window enables you to capture new video from your camcorder.

To open existing movie files or other digital images stored on your hard drive, you click the Get Movies button. You can then choose video files that you've captured earlier as a part of your movie.

In addition to the options on the toolbar, Sonic MyDVD offers one more method of modifying your movie before you create the disc. If you double-click a menu button in the project window, you will open the Trimming window, shown here:

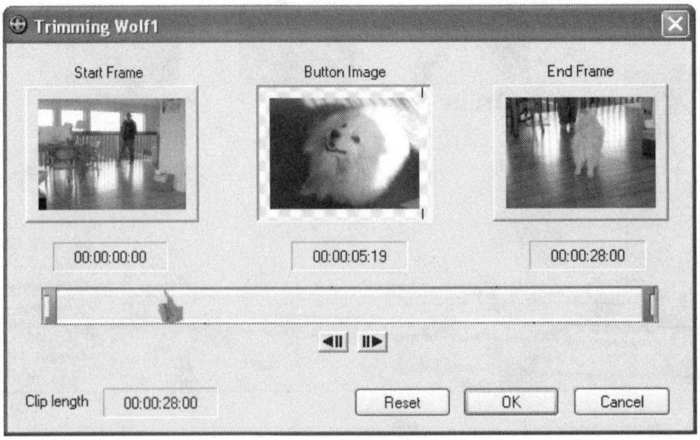

This window has three controls that you can use to modify the video clip. You can drag the hand to choose a different video frame to appear on the menu button, you can drag the green marker (to the left of the hand) to change the starting position of the video clip, and you can drag the red marker to change the ending position. I'll explain more about how this all works in Part III, when I will step you completely through the process of creating your DVD movie using Sonic MyDVD.

Finally, if you have the Video Suite version of Sonic MyDVD, you have access to another application: *ArcSoft ShowBiz*, as shown in Figure 2-5. This add-on for MyDVD provides some additional video editing options that enable you to further enhance your movies before you burn them to a disc.

> **TIP** *If your camcorder or DVD-R/RW drive came with the OEM version of Sonic MyDVD, you'll find an inexpensive upgrade to the Video Suite version available on the Sonic Web site (www.sonic.com).*

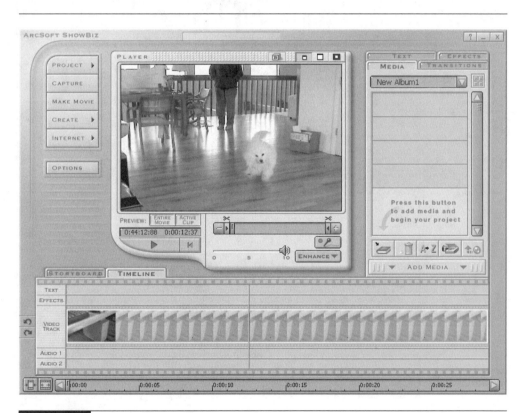

FIGURE 2-5   ArcSoft ShowBiz works with Sonic MyDVD to provide more video editing options.

Pinnacle Expression

Pinnacle Expression is an excellent choice for anyone who wants a simple, easy-to-use digital video editor and disc creator—one that has more flexibility than you might expect in a basic offering.

Pinnacle Expression maintains the ease of use that is typical of template-based digital video editors, and it also provides you with a wide assortment of options for editing your video to make it come out exactly the way you like. This benefit becomes very clear when you click the button shown in Figure 2-6 to edit your video. I opened a captured video file in Pinnacle Expression, and as Figure 2-7

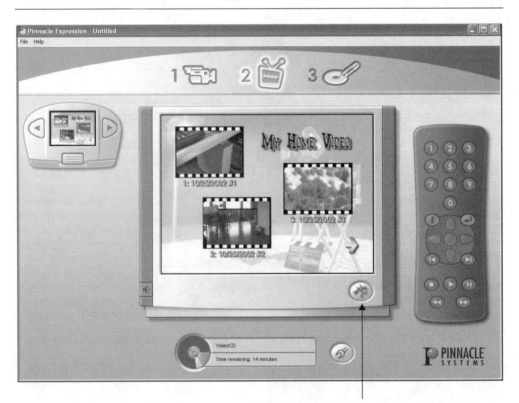

Click this button to edit your videos.

FIGURE 2-6   Pinnacle Expression is an excellent choice if you want a simple digital video editor.

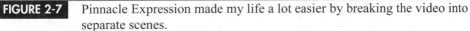

FIGURE 2-7   Pinnacle Expression made my life a lot easier by breaking the video into separate scenes.

demonstrates, Pinnacle Expression was able to break the video into 21 different scenes (using the time codes recorded by my digital camcorder). Quite clearly, it will be very easy to edit out unwanted material when the video has been broken down into the different scenes! Pinnacle Expression uses the time codes to create individual scenes automatically.

In addition to doing an excellent job of breaking the imported video into separate scenes, Pinnacle Expression also provides a number of tools that enable you to further fine-tune your movie. These include the following handy features:

■ You can combine scenes into a single scene if necessary.

■ You can break an individual scene into several scenes. This capability might be handy if a particular clip includes several useful sections, but also contains some you want to eliminate.

- You can specify an exact point where a scene should be split into two scenes.

- You can rearrange scenes by dragging and dropping them within the list of scenes.

- You can specify options for the transitions between scenes, as shown here. This feature enables you to produce a much more professional looking movie without a whole lot of extra work.

```
┌────────────────────────────────────────────────────────┐
│ Edit Options                                         [X] │
│  ┌─ For all scenes ──────────────────────────────────┐  │
│  │  ▢  Show Title When Playing                        │  │
│  │  ▢  Return to menu after playing                   │  │
│  │  ▢  Fade to black between scenes                   │  │
│  │  ▢  Dissolve between scenes                        │  │
│  └────────────────────────────────────────────────────┘  │
│              (    OK    )        (   Cancel   )          │
└────────────────────────────────────────────────────────┘
```

- You can delete scenes that you don't want included in your movie (or bring back ones you accidentally deleted in error).

Pinnacle Expression includes another very handy tool that doesn't really fit into the area of digital video editing, but which will help you make your discs look a lot more finished. That tool is a disc labeler that can use scenes from your movie to produce graphical labels to place on your discs. This sure beats the old "write on the disc with a pen" style of disc labeling most people probably use!

VideoWave Movie Creator

Roxio VideoWave Movie Creator is the third of the basic digital video editors we'll look at briefly. Roxio is a well-known company with a lot of experience in recordable disc technology—in fact, Microsoft has licensed their software as the CD-R/RW engine in Windows XP. As you will see, this experience pays off in an easy-to-use yet quite powerful program that offers some unique options.

Figure 2-8 shows how Roxio VideoWave Movie Creator looks when you first open the program. As you can see, there are a series of buttons along the left side of the screen that provide easy access to the program's functions. And, simply

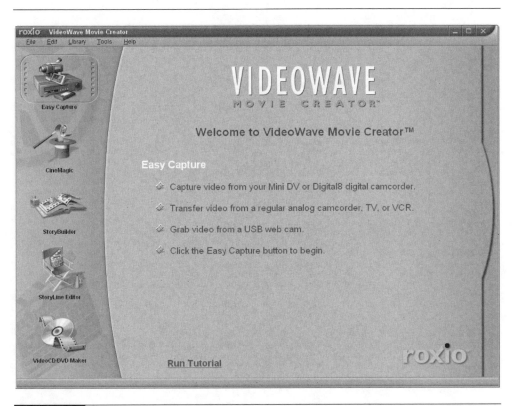

FIGURE 2-8   Roxio VideoWave Movie Creator is designed to be simple to understand.

rolling the mouse pointer over any of the buttons replaces the welcome screen with a description of what the button does. Even if you're brand-new to digital video editing, you should find Roxio VideoWave Movie Creator easy to understand.

To begin creating your movie using Roxio VideoWave Movie Creator, you click the Easy Capture button. As Figure 2-9 shows, doing this gives you the ability to choose a source for the capture (DV camcorder, analog capture through a capture card, or a USB video camera).

Roxio VideoWave Movie Creator does an excellent job of breaking down your digital video captures into scenes based on the time codes recorded by your DV camcorder. As a result, you will find that editing your movie is made easier, since you can choose individual scenes to include in your movie. This is a capability Roxio VideoWave Movie Creator shares with Pinnacle Expression as well as with the more versatile mainstream digital video editors.

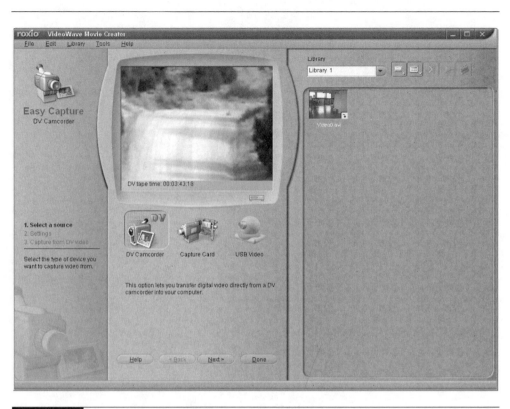

FIGURE 2-9   You can easily choose a source for your video capture.

The main Roxio VideoWave Movie Creator screen offers you four additional options. These are as follows:

- **CineMagic**   This enables you to create a simple movie that uses background music instead of narration. You might find this an excellent choice if you're in the process of converting some old silent family movies into modern DVD productions.

- **StoryBuilder**   This option opens a template-based wizard so you can create slightly more sophisticated productions using transitions between scenes, titles, and background music. This would be a great choice for getting your feet wet in digital video production without many complications.

- ■ **StoryLine Editor**   With this option you can further enhance your movies by executing even more control over the editing process. You'll want to use this option if you feel your movie needs a little more pizzazz from things like special effects and animations.

- ■ **VideoCD/DVD Maker**   Once you've finished the editing process, this option enables you to choose from a number of menu templates and to choose your final output options. This option, of course, is the one you'll need to use to complete the process of producing your DVD or VCD.

A Few Final Words on Basic Video Editors

I think the bottom line on basic digital video editors is really fairly clear. They're pretty easy to use. But even so, I'm going to recommend that you at least consider the editors I'll talk about next. For a very small additional investment, they really do offer a lot more control over the final appearance of your movie.

| NOTE | *A number of companies besides those described here have recently announced that they will soon be selling DVD creation packages that will fit into the basic video editor category. Their primary focus seems to be mostly in the area of simplifying the task of turning your home movies into DVDs. Although none of them will be available in time to include in this book, I think that it is safe to say that I feel you will end up with far superior results with one of the more full-featured editors I discuss in the next section.* |

A Look at Mainstream Video Editors

Once you decide that you want to create your own DVD movies that are a step up from the basics, you'll find that the digital video editors I like to call *mainstream video editors* are a far better choice than the basic editors we've seen so far. The reason I've decided on the term "mainstream" is that these programs really fit into a perfect niche. They're far more powerful than the entry level digital video editors, but they're also far easier to use than the pro level programs. In addition, these mainstream programs are priced far lower than the pro level editors, but they still enable you to do almost anything a nonprofessional might want to do in terms of video editing. In other words, these programs are the ones that really will fit the needs of the vast majority of camcorder owners who want to create great videos.

There are a number of very important features that typically become available to you when you step up to this level of digital video editor. These may include things like:

- More precise control over the trimming and editing of your video clips

- The option to apply transition effects selectively

- Many additional transition effect options

- The ability to add and modify soundtracks

- More menu options

- More title options

- The ability to work in different formats—for example, timelines and storyboards

- More control over the options for creating your discs

- Additional output options

Another important thing to remember about this class of digital video editors is that they're designed with the consumer in mind. That is, a lot of attention has been paid to making them easy to use in spite of all the extra features that are packed into them. This is something that I'm sure you'll come to appreciate as I give you a look at some pro level digital video editing packages later in this chapter. So let's take a look at some really great programs for creating your own DVD movies.

Pinnacle Studio

When you step up to *Pinnacle Studio*, you will really notice that this program offers many more options than are found in any of the basic digital video editing programs. For example, as Figure 2-10 shows, there is an entire dialog box in Pinnacle Studio that is dedicated to enabling you to take control of your movie from the very earliest stages. The Pinnacle Studio Setup Options dialog box enables you to control many different aspects of how the movie is captured—including how to break your movie down into scenes, what video and audio settings are used in the capture, and how music and voice are mixed with your movie—and also provides several editing options. While you don't have to use any of these options unless you want to do so, it's certainly very handy to have this level of control available.

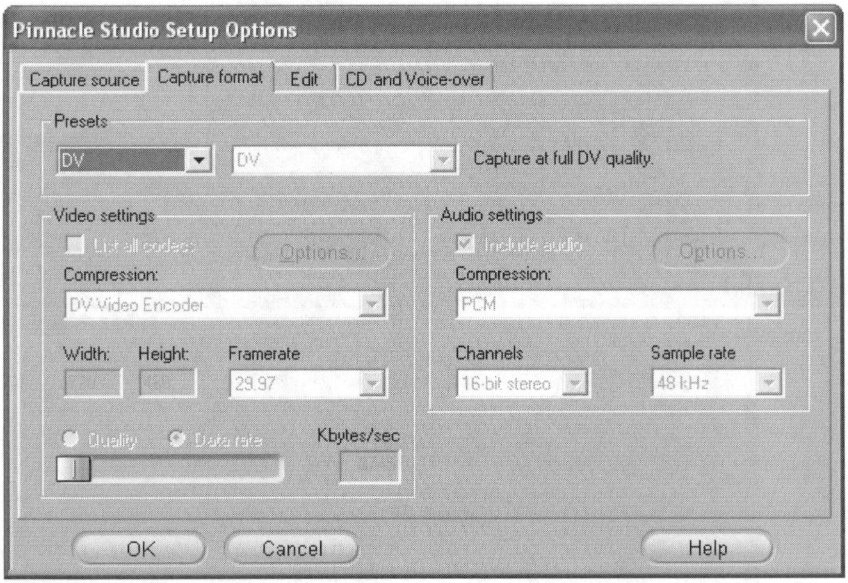

FIGURE 2-10   Pinnacle Studio enables you to customize the program's settings to best suit your needs.

One of the first things you may notice about Pinnacle Studio is that the program gives you far more tools for manipulating your movies than you get in any of the basic level digital video editing programs. For example, in Figure 2-11, I've captured a series of scenes and am now viewing my movie in the Edit panel. Along the left side of the video clip album (the area where my individual video clips appear while they're waiting to be added to the storyboard in the lower half of the window) are a number of tabs that provide quick access to various editing tools. These tabs are (top to bottom):

■ **Show Videos**   This displays all of the captured video clips in the current file using the video clip album. You can also use the drop-down list box at the top of the left-side album page to open another video so you can select additional clips to drag-and-drop into your movie.

■ **Show Transitions**   You can use this tool to add a wide range of transitions between individual scenes in your movies. You have complete control over these transition effects, and you can easily use different transitions between different sets of scenes.

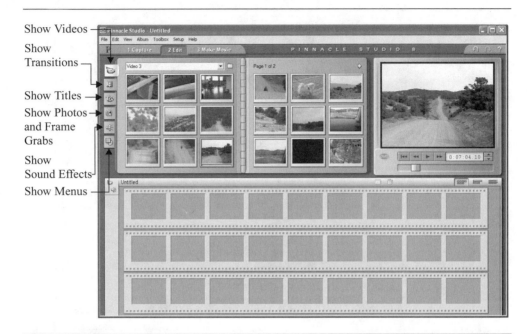

Show Videos

Show
Transitions

Show Titles

Show Photos
and Frame
Grabs

Show
Sound Effects

Show Menus

FIGURE 2-11   The tabs along the left side of the video clip album provide many very
useful editing tools.

■ **Show Titles**   With this tool you can add fancy titles anywhere in your
movie. Once you've added a title, you can easily edit it to make it fit
your movie.

■ **Show Photos and Frame Grabs**   Here you can quickly add any digital
photos or even individual movie frames to your movie. For example, you
might want to use a specific image as the background during a narration,
or you might want to use some historical photographs to create a "then
and now" view of certain scenes in your movie.

■ **Show Sound Effects**   This tool enables you to add special sound effects
to your movie. You could add your own laugh track, the sound of rainfall,
crickets chirping, or any number of other sound effects to create just the
audio effect you want.

■ **Show Menus**   With this tool you can add (and edit) DVD and VCD
menus to your project. This, of course, makes it possible for viewers
to more easily navigate to the various sections of your movie.

One place where a mainstream digital video editor like Pinnacle Studio really
shines is when you want more output options than just a standard DVD or VCD.

To illustrate this, take a look at the dialog box shown in Figure 2-12. This dialog box appears whenever you click the Settings button on the Make Movie pane. As this shows, creating a disc is just one of many different output options that are available to you. For example, you could use Pinnacle Studio to edit a movie that you could then record on your VCR, so your elderly uncle who doesn't have a DVD player could enjoy the family reunion movie. You could also create a movie clip for use on your Web site.

Sonic DVDit!

Sonic DVDit! is quite different from any of the other programs I will be showing you in this book. This difference really puts Sonic DVDit! into its own class rather than its being a part of my mainstream digital video editing group. I'm including it here because it's an important product you should know something about.

One of the biggest differences between Sonic DVDit! and the other applications in this category is that it is not intended to be a video capture application. Rather, the program uses video that you have captured using another program. For example, you might have a copy of Sonic MyDVD that came with your digital camcorder or DVD-R/ RW drive, and you would use that program's video capture options to bring

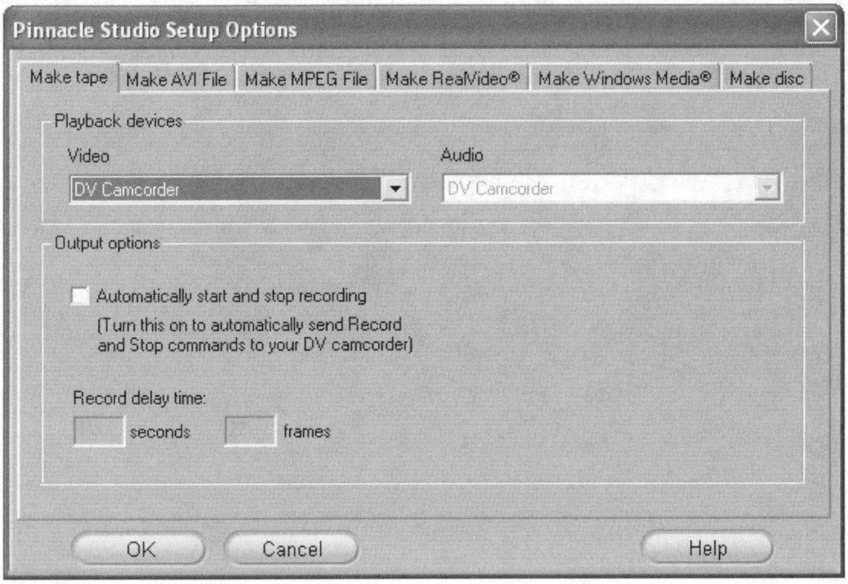

FIGURE 2-12   Tape recording is one of the additional output options available to you in Pinnacle Studio.

the video into your PC. Once the video has been captured, you could use Sonic DVDit! to add menus, titles, and audio. You can also use the program to add the *chapter points*—the breaks between scenes. Finally, you can output the movie onto a DVD or as a DVD folder.

NOTE *In discussions with Sonic, I've been told that a new version of Sonic DVDit! will be released in the near future and that this new version will include features such as video capture and additional video editing capabilities. Although the current version could be more properly called a DVD authoring program, the new version should be a better fit into the digital video editing and production arena.*

Figure 2-13 shows an example where I'm creating a movie from an existing video file in Sonic DVDit!

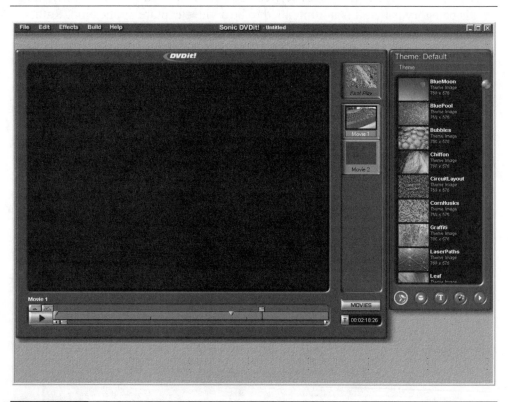

FIGURE 2-13   Sonic DVDit! focuses primarily on preparing your videos as properly formatted DVD movies.

One feature of Sonic DVDit! that you may find especially interesting is the extreme flexibility the program offers in creating DVD menus. In Sonic DVDit! you build your menus manually by creating the buttons you want and then linking them to specific chapter points. You can have multiple menus, and you can have up to 36 menu buttons. When you burn your DVD, Sonic DVDit! automatically formats the menus so that they will be compatible with virtually all set-top DVD players.

Roxio VideoWave Power Edition

Roxio VideoWave Power Edition is another excellent choice in a mainstream digital video editing and DVD production program. Like Pinnacle Studio, Roxio VideoWave Power Edition offers the full range of features, from video capture through a complete range of editing options, and offers a wide range of output options to meet your needs. Figure 2-14 shows where I'm just beginning to edit a movie in Roxio VideoWave Power Edition.

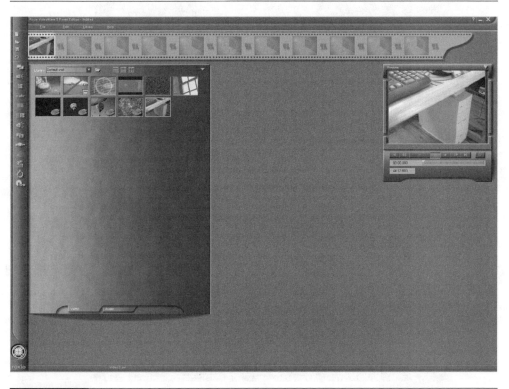

FIGURE 2-14   Roxio VideoWave Power Edition offers a full range of video editing tools.

One of the really cool features in Roxio VideoWave Power Edition is something called *TimeWarp* (see Figure 2-15). Although a number of digital video editors have the ability to play scenes faster or slower than normal, TimeWarp actually produces a new video clip that has been altered to look far better than a standard video clip would when played at a different speed. For example, if you ask TimeWarp to create a slow-motion clip, it can actually produce a smooth-motion effect instead of the jumpy one you might see if you play a standard video clip slower than normal.

A Few Final Words on Mainstream Video Editors

Where the basic digital video editors offer simplicity and ease of use, the mainstream editors offer considerably more power and additional options for creating a far more

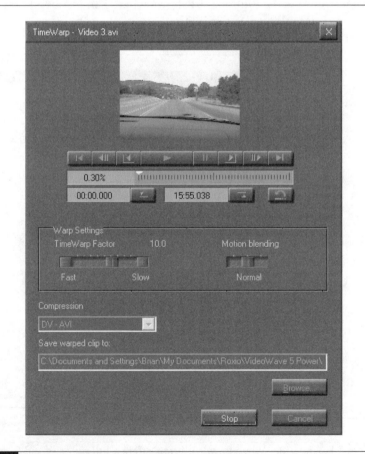

FIGURE 2-15 The TimeWarp feature produces superior results when you want to vary the playback speed of a video clip.

personalized movie. The mainstream editors do take just a bit more effort, of course, but they really do give you a lot more, too.

If you want a complete digital video editing package, you certainly can't go wrong with either Pinnacle Studio or VideoWave Power Edition. For additional authoring capabilities, Sonic DVDit! adds enough features to make it a full-featured addition to this class, too.

A Glance at Pro Level Video Editors

One of the biggest problems for most people in terms of choosing the correct software for their needs is simply that there are so many choices and so little good advice available. The situation is really no different when it comes to choosing a digital video editor. That's one reason why I'm going to take a few pages to introduce you to what I call *pro level* video editors.

The pro level editors are all great programs. They're also very expensive—in fact, most of them probably cost more than what you likely would pay for a very nice, brand-new digital camcorder. These programs have a lot of power, but it's likely no one would call them easy to use. In fact, all of them are downright complicated. Believe me, these are programs that are not intended for anyone but a professional video producer!

So why am I showing you these pro level video editors? I'm doing it for one simple reason. At some point, I'm sure that someone will tell you that you just can't live without one of these programs, that they're absolutely essential if you want to create your own DVD movies. By introducing you to these programs now, I'm giving you some ammunition to help you resist that kind of pushy person. Once you've seen these editors in the following pages, you'll honestly be able to say that you know about them and that you simply don't need that much power to do the sort of video editing you intend to do.

Of course, if you are planning on going into professional video editing, maybe one of these programs will be right for you. But take the time to get your feet wet with something a bit simpler—you'll probably be a lot happier in the end.

Adobe Premiere

If someone just insists that you have to use a particular video editing program, the likelihood is very high that the program they're pushing will be *Adobe Premiere* (see Figure 2-16). It is probably fair to say that more people have heard of Adobe Premiere than of any of the other video editors I discuss in this book.

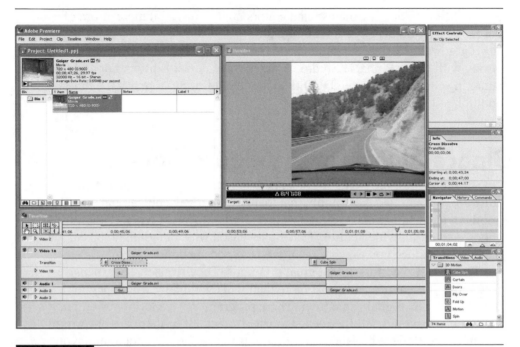

FIGURE 2-16   Adobe Premiere is certainly an excellent pro level video editor.

There is virtually nothing in the video editing arena that you cannot do in Adobe Premiere. For example, in Figure 2-17 I'm in the process of designing a fancy transition effect to fit between two scenes in my movie.

While it's certainly true that Premiere has tons of power, notice how complicated the program looks in the two figures. This isn't a program that you'll install on your PC and turn out your first finished DVD in an hour or two. In fact, the Adobe Premiere package includes a one-hour orientation video!

Pinnacle Edition

Pinnacle Edition is the next pro level video editor we'll look at (see Figure 2-18). Here, too, is a very powerful tool that is really far more than what an average user needs. In fact, Pinnacle Edition would likely seem pretty confusing to a casual movie maker, since it essentially throws away all vestiges of the familiar Windows interface and replaces it with tools designed specifically to make digital video editing tasks easier—for the pro level user, that is.

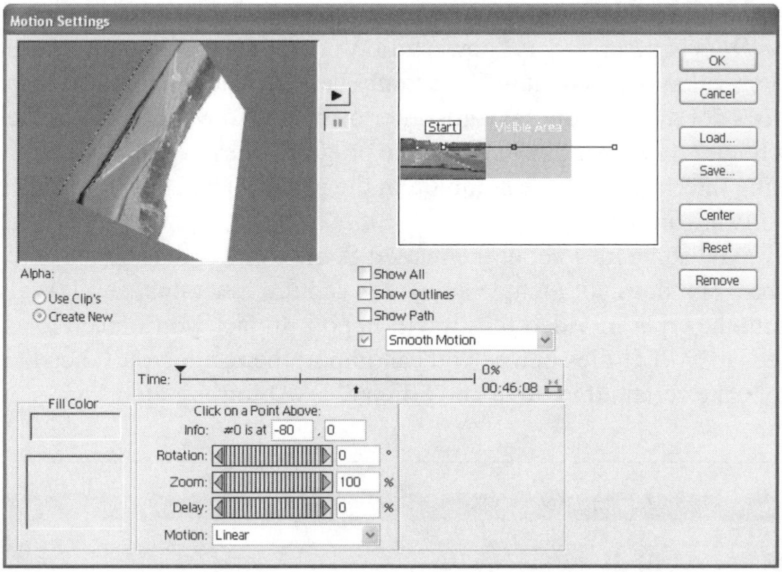

FIGURE 2-17   Fancy transitions are just one aspect of the power of Adobe Premiere.

FIGURE 2-18   Pinnacle Edition turns your PC into a video editing workstation with its own appearance.

Sonic ReelDVD

Sonic ReelDVD is described as "powerful DVD production for corporate and independent video professionals." I certainly believe that description—especially since it does not mention digital camcorder owners who want to create their own DVDs at home. As Figure 2-19 shows, the program doesn't sport one of those intimidating interfaces that are common in the pro level category, but this is some serious software anyway!

To give you some idea about the market Sonic ReelDVD is aimed at, consider this: the program does not include any video capture functions, and it is also very picky about the types of video files it will import. In fact, you'll need to convert your videos to MPEG files before you can import them, and you'll need to be very careful to make certain they're in an "official" DVD format, too.

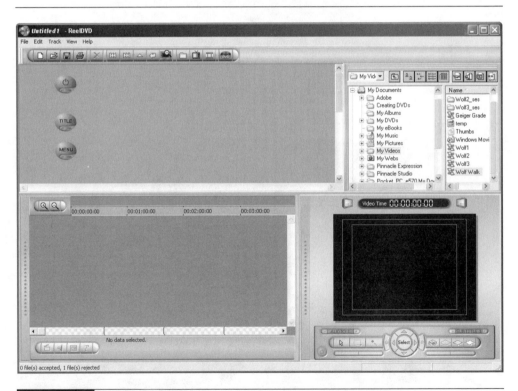

FIGURE 2-19   Sonic ReelDVD hides a lot of complexity under a simple appearance.

Sonic DVD Producer

The final pro level editor I'll show you briefly is *Sonic DVD Producer* (see Figure 2-20). This is a program aimed squarely at the video professional.

FIGURE 2-20   Sonic DVD Producer is a true pro level DVD production tool.

It includes such features as the ability to include up to eight different audio streams and up to 32 different subtitle tracks. In other words, this is the program you would want to use if you were going to produce videos with foreign language dubbing and the viewer's choice of foreign language subtitles.

Both Sonic DVD Producer and Sonic ReelDVD have heavy-duty copy protection in the form of a *dongle* that you must plug into the parallel port on your PC in order to run the programs. Needless to say, these people are serious about making certain that only authorized users can access the programs.

The Final Word on Pro Level Editors

There's no getting around the fact that all of the pro level editors I teased you with are extremely powerful (and expensive) programs. But there's also no getting around the fact that these programs really aren't appropriate for most people who have a camcorder or a bunch of old home movies on tape and just want to create their own DVDs.

Well, now that I've given you a quick look at the range of video editing options that are available, let's move on and take a look at how you can get your videos into your PC so that you can begin having some fun.

CHAPTER 3

Capturing Your Video

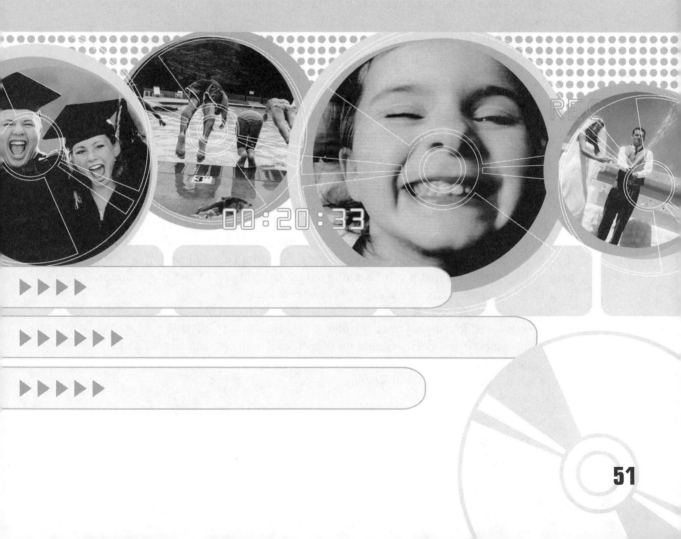

▶▶▶▶▶

▶▶▶▶▶▶▶

▶▶▶▶▶▶

Now that you've had a very brief introduction to some of your choices in digital video editors, I'd like to turn to another area that may be causing you some confusion. That's the procedure for actually getting the content for your movies from your camcorder, VCR, audio CDs, and even old photographs into your PC. This is, of course, a very important step in the whole process, because you can't really do much in the way of editing until you have something to work with.

In this chapter, we're going to consider the hardware options you can use to transfer various types of content into your PC. Some of these may already be a part of your computer—but you might not realize how to use them (or even their purpose) at this point. It's also possible that you'll have to install an adapter card in order to be able to transfer the content into your PC. Whatever the situation, you'll find that setting up your PC is not difficult at all. And even if you don't want to do the work yourself, knowing what is required will make it a lot easier when you ask someone to assist you.

So let's begin by looking at what you need in order to move your content into your PC so that you can begin having some fun.

Understanding Video Sources

I'd like to start out by making certain that you understand a little bit of the basics of video sources. Essentially, there are only two types of video: analog and digital. *Analog video* is the type of video most TVs play, and there are several sources that you'll find quite familiar, including over-the-air TV signals, VCRs, older camcorders, and so on. *Digital video* is the type of video produced by digital camcorders, and it is what is recorded on DVDs.

Eventually, analog TVs are supposed to go away and all TV broadcasts will be digital. For now, though, all digital video is converted into an analog signal by the playback apparatus so that it can be displayed on an ordinary TV set. For example, your DVD player outputs an analog signal to your TV.

Inside a computer, on the other hand, all signals are digital. Any analog signals must be converted into digital ones in order to be edited in a PC. That's because PCs are digital computers, and they cannot work with analog data—only digital data.

Digital data storage offers a number of advantages over analog data storage. These advantages include the following:

- Digital data can be stored in much less space than analog data.

- Digital data can be copied as many times as you like without reducing the quality.

- Digital data is generally considered to be more stable and therefore offers a longer lifespan than analog data.

- Digital sources typically produce higher quality output than analog sources.

- Digital data can be manipulated by computers (the most significant advantage for our discussion).

The most important things for you to remember here are that analog and digital video sources require different methods to move the data into your PC and that, once that data is in your PC, there's no essential difference between the two.

Using DV Camcorders

Of all of the possible sources for your content, a *DV camcorder* (digital video camcorder) like the one shown here is probably the best overall choice:

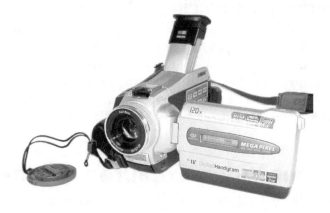

Digital camcorders typically offer superior image quality compared to the older analog ones, but the one factor that makes DV camcorders so great for digital video editing is simply that (in most cases) your PC can control the playback of a DV camcorder.

Having your PC in control of your DV camcorder means that you can use your PC to start, stop, rewind, fast-forward, or even advance or rewind just a frame or two. As a result, it's a bit easier to capture exactly the scene you're interested in adding to your movie.

DV camcorders do offer one additional and very important feature not found in analog video sources. DV camcorders automatically record a *time code* along with your video. Most digital video editing programs can use this time code to automatically detect the scenes in your movie. Whenever the time code is discontinuous, your DV camcorder is stopped, so it's pretty likely that this is the perfect place to declare that a new scene has begun.

Using Analog Video

Non-digital camcorders and VCRs produce an analog signal that must be converted to digital before it can be used in your PC. Although the picture quality of analog sources is typically lower than that of digital ones, I'm sure that most people would agree that being able to save your old family movies onto the more modern DVD is more important than quality that might not be quite as good.

When you use an analog video source, your PC can capture the video, but it cannot control the playback. Instead, you must start and stop the VCR or camcorder manually. It's usually best to start the playback a bit before the point where you want to begin capturing the analog video and to continue slightly past where you want to stop the capture. That way, you'll have a little room to cleanly edit the video clip to get everything you want.

Using Existing Video Files

In addition to capturing new video for your projects, you can also add existing video files to your DVD movies. Depending on the digital video editor you use, you may be able to use video files in a number of different formats. Typically, most of the mainstream digital video editors will allow you to use *AVI*—audio video interleave—and *MPEG*—Motion Picture Experts Group—files. Your favorite editor may even accept a few other video file formats, but AVI and MPEG files are usually a safe bet.

A Look at Your Video Capture Options

Before you can edit your videos on your PC, you must first get them into your PC. This requires specific types of video capture hardware that matches the source that you intend to use. In this section, we'll take a look at the specifics of video capture hardware.

FireWire (IEEE-1394)

If you use a digital camcorder, it almost certainly uses a type of connection variously known as FireWire, IEEE-1394, or i.Link—these are simply three names for exactly the same connection (which I'll call IEEE-1394 from now on). This is a high-speed data connection that is perfect for transferring digital video into a PC.

Some PCs come equipped with IEEE-1394 ports, and if your PC has them, you won't have to add anything else in order to connect your digital camcorder. You can look in your system's documentation, or you can simply look for connectors that look like the ones shown to the left.

If your PC doesn't have any IEEE-1394 ports, you'll need to add an adapter board like this one which comes inside the Pinnacle Studio Deluxe box:

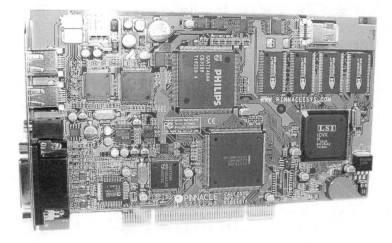

Adding an adapter board is a fairly simple process, but you do have to turn off your computer, open the case, locate an empty PCI slot, remove a blanking plate, and insert the board. It is absolutely essential that you take care not to subject the board to any static electricity, which can easily damage the board; if you're not comfortable installing the board yourself, ask a knowledgeable friend for a little assistance.

You will also need an IEEE-1394 cable like the one shown here (also included with Pinnacle Studio Deluxe):

Notice that one end of the cable is smaller than the other. The small end plugs into your digital camcorder, the large end into your PC.

Analog Video

If you want to use an analog video source, you'll need a video capture board—something that is not found in most PCs, even those equipped with an IEEE-1394 port. A video capture board converts the analog signal into digital data, which can then be processed by your PC. The adapter board included in Pinnacle Studio Deluxe also serves as an analog video capture board, and it includes a connection box, which looks like this:

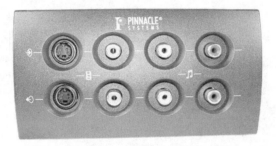

This connection box has several connectors. These include the following:

- Composite video input to move the video signal into your PC

- Composite video output so you can record your finished movie on your VCR

- Right and left channel audio inputs to copy the stereo audio into your PC

- Audio outputs so you can record the sound on your VCR

- S-Video input to copy a higher quality video signal into your PC

- S-Video output to record your movie on an S-Video VCR

If you use the composite video from your VCR or analog camcorder, you'll need an A/V cable like this one:

These cables are al[...]:d for right channel
audio, and white f[...] camcorder has an
S-Video output, y[...]d you'll need to use
the red and white [...]

> **TIP** *If you ha[...]ource, using it will*
> *improve [...]forget to connect the*
> *audio sig[...]any audio information.*

RF (TV Tu[...]

If you happen t[...]es not offer either
composite or S[...]e to add content from
your existing [...]st. You can simply add
a TV tuner ca[...]own here:

As an added benefit, adding a TV tuner card will enable you to record TV shows
onto your hard drive or even onto a DVD. You should, of course, be aware that TV
broadcasts are copyrighted material, so you cannot include any such content on a
DVD you intend to sell (unless you obtain written permission from the copyright
holder).

> **TIP** *If you want to learn more about putting together your own PC that has all of*
> *the features you might want for video editing, check out* Build Your Own PC
> Home Entertainment System, *published by McGraw-Hill/Osborne Media*
> *Group (and written by your favorite technology author, Brian Underdahl).*

Understanding Your Audio Sources

Have you ever watched a truly silent movie? It's pretty unlikely that you have, since even the old "silent" films from the early days of Hollywood were usually viewed with the accompaniment of live music that was played in the theater. The reason for this was pretty simple: life is full of sounds, and movies seem far more realistic if they include a soundtrack. In this section we'll take a quick look at some of your options for creating a soundtrack for your movies.

Using the Recorded Soundtrack

When you record a video on your camcorder, the camcorder also records a soundtrack, which is synchronized with the video. In the case of a very simple movie, this may be all the soundtrack you need. You may find, however, that the recorded soundtrack really doesn't fit your needs. Here are some typical problems with recorded soundtracks:

■ When you film a video with your camcorder, the microphone tends to pick up all of the sounds—including the directions you give to someone from behind the camera.

■ Even when the soundtrack is fairly clean, you may need to edit some scenes to produce a better movie. If you do, the recorded soundtrack will be cut, too, unless you apply some advanced sound editing techniques (see Chapter 6).

■ There may simply not be enough sound worth recording. For example, if you film a sequence while you're riding in a car, the soundtrack probably won't pick up much beyond wind and road noise.

Even with all of the problems you may encounter with the recorded soundtrack from your camcorder, it will probably be a major element in most of your movies. Fortunately, it doesn't have to be the only one.

Adding Music from Audio CDs

Another potential source of audio for your movies is the music on audio CDs. Your PC can easily copy music from audio CDs, and you can then add that as background music. In fact, as you'll see in Chapter 6, your digital video editor may even have this function built into the program as an option. There is, however, just one problem with this scenario—the music on audio CDs is protected by copyright, making it illegal for you to use most recorded music this way.

From a practical standpoint, it's unlikely that the music police are going to come and beat down your doors if you use a cut from your favorite audio CD as background music for a DVD movie you create for your own personal use. It's a very different matter, however, if you are producing a DVD for any type of commercial purpose. In that case it is far better to consider the option I'll show you next.

Using Music Samples

Rather than taking the risk of using copyrighted music in your movie, you may find this next option somewhat more appealing. Some video editors include stock music that you are free to use in any movies that you create with the editor. For example, Figure 3-1 shows the SmartSound pane in Pinnacle Studio. This pane enables you to add music at various points in your movie without having to obtain permission from the musicians and without having to pay any fees.

When you add stock music to your movies, you typically have the option of deciding how long the music should play. You can include anything from very short

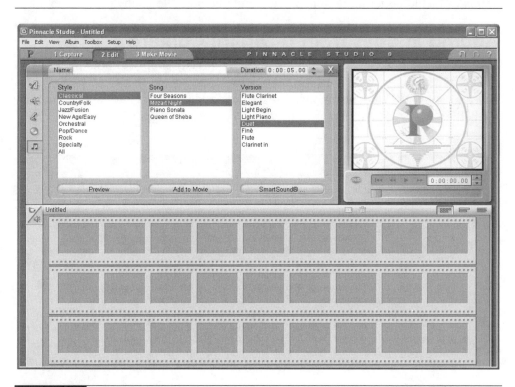

FIGURE 3-1   Using stock music eliminates the problem of copyright infringement.

little pieces of music to longer segments that accompany a good part of your movie. In fact, that's why Pinnacle calls this feature SmartSound—it is so adaptable.

Adding Sound Effects

You may also want to add sound effects to your movies. In a commercial motion picture, these are created by someone known as a *Foley artist,* but you may find that your digital video editor offers a whole range of sound effects you can add yourself, so you won't have to hire someone to do that for you.

Figure 3-2 shows just a few of the sound effects that are built into Pinnacle Studio. These sound effects can really liven up a movie!

Using Other Content

So far we've been concentrating on moving images and sounds, but these aren't the only things you can add to your DVD movies. In fact, a DVD or VCD can be an excellent method of distributing your family photo album, because you have the option of including far more things than would be possible in any printed version.

Adding Images from a Digital Camera

If you have a digital camera, you probably have a ready supply of photographs of family gatherings, vacations, birthday parties, and so on. These can easily be added to your DVD movies. In fact, most digital video editors have a very simple method of adding digital photos to the *storyboard* (or *timeline*).

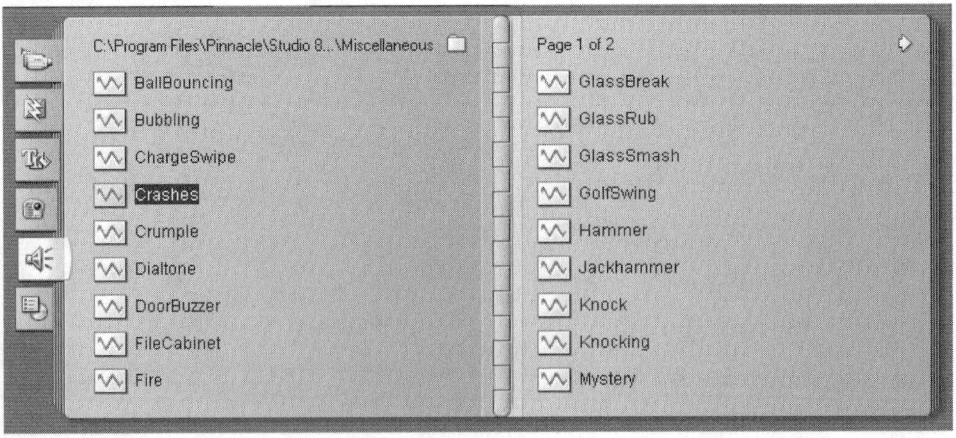

FIGURE 3-2   You can add sound effects to your movies just by choosing from a selection of recorded sounds.

When you add digital images to your movie, you'll want those images to display for a long enough time so that viewers can see the images. You'll probably also want to add background music or even a narration track. And don't forget that it's perfectly acceptable to mix still images in with video segments.

Scanning Old Photos

You can also add scanned images to your movies just as easily as you can add images from a digital camera. This would be an excellent way to include photos from your old family album, or you could use some historical images to set the scene for a "before and after" view if you're creating a travel video.

Scanned copies of old documents or greeting cards are also items that you may want to consider including in your movie. I can imagine creating a family album that includes letters and postcards someone sent home from a trip abroad. You could also include things like a scanned copy of an antique greeting card to set the mood for a holiday video.

In this chapter, I've tried to give you the information about how you can get your content into your PC as well as ideas for some things you might want to add to your movies. As you've seen, the distinction between digital and analog content only exists outside of your computer—once the content is inside your system, it is all handled as digital data.

Next, we're going to look at some important movie-making concepts that you need to understand in order to get the most from your digital video editor. Once you have learned these fundamentals, you'll find that you have a lot more fun creating your great-looking movies!

PART II 00:20:33

Movie Making Concepts

CHAPTER 4

00:20:33

Storyboards and Timelines

▶▶▶▶▶

▶▶▶▶▶▶▶

▶▶▶▶▶▶

It's time to move on to some important movie making concepts. These are really the fundamentals that you will want to understand in order to get the most from your video editing software. By taking the time to learn these basics, you will be able to produce movies that are better looking and a lot more enjoyable to watch.

In this chapter we're going to look at *storyboards* and *timelines*—two different but related means of visualizing the flow of the scenes that make up your movie. As you will see, storyboards and timelines also provide you with the ultimate control over just how all of those scenes fit together and, consequently, over the overall flavor of your movie. In fact, as you increase the sophistication of your productions in later chapters, you'll find that these two objects are really the backbone of the entire movie making process.

Before we go any further, I'd like to take this opportunity to point out that in this section of the book we'll be using one digital video editor—Pinnacle Studio 8—as the basis for our examples. The reason is simply that Pinnacle Studio 8 is one of the most complete of the mainstream video editors—one that can demonstrate all of the concepts for you—its consistent use will reduce any possible confusion. Rather than trying to understand a whole bunch of different programs, you can simply spend the time learning about the important movie making concepts, which you can then apply in your favorite video editor.

Remember, though, that even if you prefer to use a different video editor, I'm not going to ignore your needs. In fact, in Part III I'm going to devote each chapter to creating a complete movie in a different popular video editing program. Chances are, you'll find a whole chapter on your favorite. Until then, however, let's get some basics under your belt so you can be a master movie producer!

Understanding Storyboards

A storyboard is one of the most basic elements of movie making, and it's also probably one of the most intuitive ones. As soon as you see a storyboard in use, your reaction is almost certain to be something on the order of "well, of course." The whole idea behind a storyboard is simply that it shows the layout of the scenes in the order they'll appear in the movie. For example, as shown here, I've begun laying out a movie that at the time has five scenes.

When you look at the storyboard, it's pretty easy to tell that scene 2 will play before scene 3 begins and that both of them will play before scenes 4 and 5. As you will soon see, this does not necessarily mean that the scenes were originally shot in the order in which they appear on the storyboard.

How Storyboards Work

Digital video editing is a very visually oriented affair. When you create the storyboard for a movie in a digital video editor, the storyboard actually shows the beginning frame of each scene, so that you can tell at a glance which scene is which. In traditional movie making, the storyboard was usually created before the movie was filmed, and the scenes were represented by hand-drawn images. Even so, the general concept remains pretty much the same, and the storyboard shows how the scenes will flow in the final movie.

By the way, there is no reason why you cannot create an old-fashioned (traditional) storyboard as a part of your movie making process. In fact, you'll probably want to do so if your goal is to create a movie that tells a particular story or that serves a specific purpose. For example, if you are creating a recruitment video for your club, you'll probably want to show prospective members a number of interesting and fun activities in which the club members participate. By creating your own storyboard, you can be sure to film all of the scenes that you'll want in your final movie. Of course, it's not really necessary for you to do something as formal as actual drawings of the various scenes—you might find that simply creating a list of the things you want to film will be enough. Then you can just check the scenes off your list as they're filmed and move on to the next item on the list.

The storyboards in digital video editors all have an appearance similar to a filmstrip. Whenever you add a scene to the storyboard, you increase the length of the filmstrip by one scene. In every case, the scenes will all appear adjacent to each other. That is, there can be no empty spots between scenes. If you delete a scene from the storyboard, the remaining scenes will shift over to fill the gap.

Adding Content to Storyboards

So just how do you add the scenes to the storyboard? Actually, adding scenes is very easy: you just click on a scene in the album, hold down the left mouse button, drag the scene onto the storyboard, and release the mouse button. For example, here I'm dragging a new scene onto the storyboard and am about to drop it into the sixth position on the storyboard.

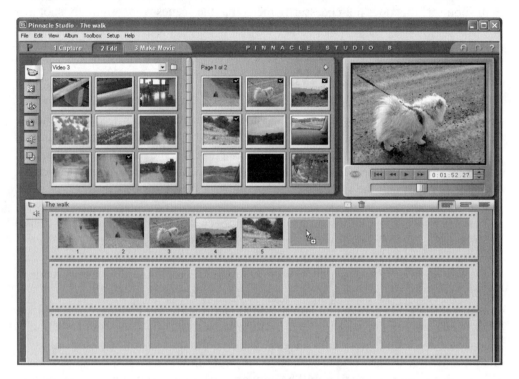

Once I release the mouse button, the first frame of the new scene will appear on the storyboard, but until then only the mouse pointer and a highlight frame show where the scene will be.

TIP *Scenes are also often referred to as "clips."*

In addition to adding clips to the end of the storyboard, you can also add them to any other position within the existing scenes. To do so, simply drag and drop the new scene between two existing clips. When you do, the scenes to the right will automatically shift one frame to the right to make room for your added scene.

It's important to realize that you don't have to include all of the scenes that are contained in the scene album in your movie. In fact, an important part of the editing process is deciding which scenes should be left out. Remember: it's far better to leave out a boring scene or one where you flubbed the filming than to include it simply because it's there. Your audiences will thank you for being selective!

Moving Scenes on the Storyboard

As you add scenes to your movie you will no doubt discover that sometimes a scene just doesn't seem to be in the correct location. Perhaps you have someone opening a door to welcome a visitor in one scene, and then the next scene shows the visitor knocking on the door. Clearly this is not the way the scenes should be laid out in order to tell the story in a linear, logical fashion.

Moving scenes on the storyboard is a simple drag-and-drop process. First you select the scene by clicking on it, and then you drag it to its new location. When you drop the scene, the remaining scenes will shift left or right as necessary to accommodate the new scene order.

> **TIP** *To remove a scene from the storyboard, select the scene and then click the trashcan icon. Different editors place this icon in different locations—Pinnacle Studio displays the trashcan icon just above the top row of the storyboard's filmstrip—but they all seem to use a similar icon for deleting scenes.*

Editing Scenes on the Storyboard

Sometimes you'll find that you want to use only part of a scene rather than the whole scene in your movie. You might find that the first section of a scene begins too early, that part of the scene is too bumpy, or maybe that the end of the scene just goes on too long. Whatever the reason, you may need to edit the scene by trimming it.

> **NOTE** *Unfortunately, not all digital video editors provide you with the tools necessary to trim scenes after you have captured them. With some editors your only option is to break the video into short clips during the capture process and then add the clips you want onto the storyboard. Although this is not nearly as convenient as being able to trim the clips after capture, it does provide you with a method of eliminating unwanted footage from your movie.*

Figure 4-1 demonstrates how you can edit a clip in Pinnacle Studio. In this case, I want to remove some unnecessary material from the beginning of the scene. Let's take a closer look at the process of trimming a scene:

1. First, select the scene you wish to edit in the storyboard.

2. Next, click the video toolbox icon to open the toolbox (this is the area above the storyboard that replaced the scene album). You can also open the toolbox by double-clicking the scene.

3. To trim material from the beginning of the scene, drag the left-most marker until the viewing window shows the frame where you want the scene to begin.

4. To trim material from the end of the scene, drag the right-most marker.

Click here to open the video toolbox.   Drag this marker to set the starting point of the scene.   Drag this marker to set the ending point of the scene.   List view

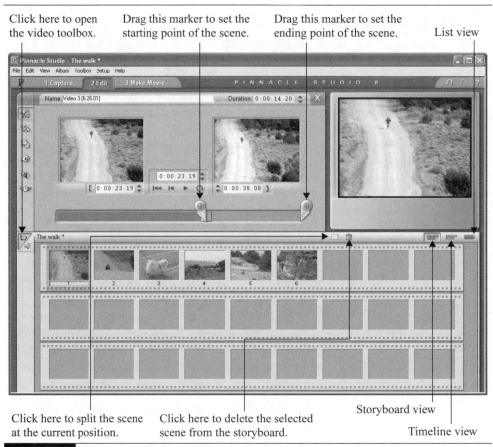

Click here to split the scene at the current position.   Click here to delete the selected scene from the storyboard.

Storyboard view

Timeline view

FIGURE 4-1   You can trim the beginning or ending of a clip so that you include just the material you really want.

You can also split a scene into two scenes by advancing the player to the point where you want to split the scene and then clicking the razor blade icon. This can be handy when you want to cut out material from the middle of the scene, since each new scene can then be trimmed individually. You can continue splitting the resulting scenes into more scenes if necessary.

Understanding Timelines

At first glance, a timeline doesn't look all that different from a storyboard (see Figure 4-2). A storyboard and a timeline are similar in a number of important ways. They both show the scenes in the order they'll appear in your movie, they

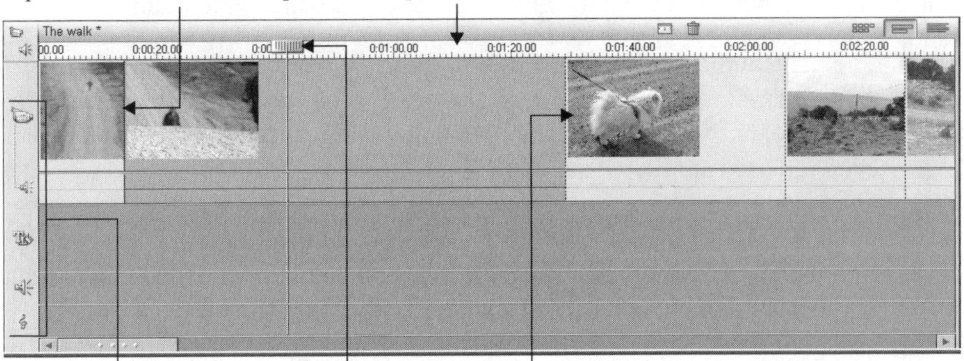

Drag the beginning of a selected
clip to trim the start of the clip.

Drag in the time scale to
expand or contract the scale.

Click to lock the main
video track, main audio
track, title track, sound
effect and voice-over track,
or background music track.

Drag to view
different frames.

Drag the end of a selected clip
to trim the end of the clip.

FIGURE 4-2   The timeline view offers another way to work with the scenes in your movie.

both support drag-and-drop placement of new scenes, and they both allow you to
rearrange scenes by dragging and dropping.

How Timelines Work

If you take a closer look at the timeline view, you'll soon notice that it does have
a number of important differences from the storyboard view. The most obvious of
these differences is that the timeline actually indicates the length of time that each
scene will play relative to all of the other scenes. That is, the amount of space a
scene occupies on the timeline is a proportional representation of its length in the
movie. And, of course, the time scale shows the actual run time in hours, minutes,
seconds, and hundredths of seconds.

NOTE
*Your first look at a timeline may be somewhat confusing, since it appears
as though there is blank space between scenes. In reality, this is not the case.
Rather, each scene is represented by a thumbnail view of the first frame of
the scene followed by as much blank space as is necessary to indicate the
running time of the scene. In some cases, the scene may actually be short
enough so that only a small sliver of the thumbnail is visible.*

As Figure 4-2 showed, the timeline has additional tracks beyond the simple video track that shows the recorded video. I'll discuss these additional tracks in more detail in later chapters, but for now let me give you a brief introduction to them:

■ The main audio track contains the sounds that were recorded by your camcorder along with the video. This track is normally synchronized with the video, but you can separate it so that it can be used independently (or even not used at all).

■ The title track is used to display titles that are superimposed over the video.

■ The sound effect and voice-over track is another audio track that you can add if you want to include recorded sound effects or narration in your movie.

■ The background music track enables you to add music, of course.

You may well wonder why you might ever want so many different audio tracks. The answer is simple: by keeping the different audio tracks separate, you can control their volume independently. I'll explain more about using audio tracks in Chapter 6.

Adjusting Clips on the Timeline

Although a timeline might seem like a complicated and very technical type of tool, it is actually very simple to use and highly intuitive. If you'll recall from Figure 4-1, trimming a clip in the storyboard view requires you to open a separate video toolbox so that you can adjust the beginning or ending of the clip. When you're working in the timeline, the process is much more straightforward—you simply drag the beginning or ending of a clip to trim it. For example, here I've dragged the beginning of the second clip to the right to trim off a bunch of frames:

As a result, the overall length of the movie is now shorter, and the scenes following scene 2 have shifted to the left to fill in the gap.

Before you can adjust a clip on the timeline, make certain that you have selected the clip. The selected clip is always shown with a different background color to make it easier for you to identify which clip is selected. You can adjust a clip only if it is selected.

As you adjust the length of clips, you'll probably find on occasion that you have trimmed off a bit too much from the beginning or the ending of the scene. If so, simply drag the end of the clip that you want to tweak in the opposite direction. For even finer control, hold the mouse pointer over the time scale and drag to the right to expand the scale. You can always drag the time scale to the left when you want to see more of your movie at once. Changing the scope of the time scale does not affect the length of the movie, but only the fine-level control over the beginning and ending of clips.

Before I go any further I'd like to introduce you to one of my secret weapons for digital video editing, the *Contour ShuttlePRO* seen here:

This is a special controller with 13 programmable buttons, a *jog wheel* for easily moving forward or backwards frame-by-frame, and a *shuttle ring* for variable-speed playback. If you want to make your digital video editing a lot easier, there's simply no better tool that I've ever seen for this task. You can find out more about the Contour ShuttlePRO at www.contourdesign.com.

Slow Motion and Fast Motion Effects

Movies are made up of a series of still images—*frames*—which are played back rapidly enough so that we perceive those images as portraying actual motion rather than as a bunch of individual still images. In fact, we probably do see the individual images, but because of the speed at which they're shown, we think we're seeing continuous motion.

Normally we view videos at about 30 frames per second (30 fps). Depending on the video standard we're using, the speed may range from 25 to 30 fps, but the effect is pretty much the same—as long as the frames are played back at the same rate at which they were captured, that is. It's when we vary the playback speed that things get interesting.

Imagine a scenario where you filmed a golfer at normal speed taking a swing at the ball. Now suppose that during playback you viewed the frames at a much slower rate, such as 15 fps. The result would be that the swing would appear to take twice as long, and it would seem like the golfer was moving in slow motion. The opposite effect could be used to make it look like someone suddenly took up speed walking. I'm sure you can imagine all sorts of other interesting and humorous ways to use these slow motion and fast motion effects in your movies.

You can adjust the playback speed using the playback speed tools from the video toolbox (see Figure 4-3). Let's take a quick look at the specifics.

1. Select the clip you want to adjust.

2. If it is not already open, open the video toolbox by clicking its icon.

3. Click the playback speed tools icon along the left side of the video toolbox to display these tools.

4. Drag the speed slider left for slow-motion or right for fast-motion. You'll probably want to preview the effect in the view window to the right of the toolbox.

5. If you are selecting a playback speed slower than normal, you'll probably want to make certain that the "Smooth motion between frames" checkbox is selected. This will generate *interpolated* frames, which will make any movement seem more natural.

6. To create a stroboscopic effect, drag the strobe slider to the right. This causes frames to be repeated, producing a very unnatural effect (which can be quite humorous).

Needless to say, motion effects should probably be used sparingly, because otherwise they can easily distract from the overall story. It's also important to remember that the recorded soundtrack won't play if you've changed the playback speed (unless you reset the speed to normal). In Chapter 6, I'll show you how to work around this problem.

Click here to open the
playback speed tools.

Drag to add
strobe effects.

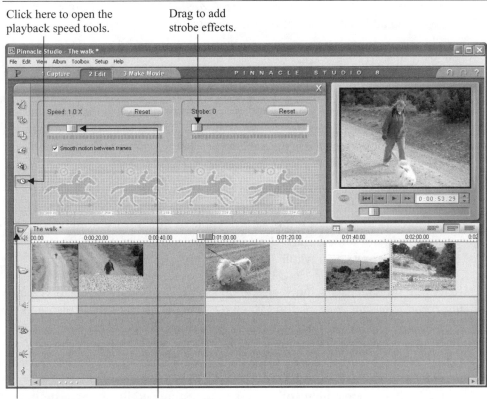

Click here to open
the video toolbox.

Drag to adjust the
playback speed.

FIGURE 4-3   You can vary the playback speed properties to create slow-motion or
fast-motion effects.

Understanding Scenes

The final item I'll discuss in this chapter is one that you probably feel you already
know about. In fact, you already do know quite a bit about *scenes* even if you're
just starting with digital video editing. But there are still some things you may
still need to learn about this subject if you really want to make the best possible
movies. For example:

■   Scenes don't have to be discontinuous. If you split an existing scene into
two parts and then play the movie without doing any additional editing,
those two scenes will play back just as if they were a single scene. This
characteristic may become more important as you start playing around

with special effects, such as *fades* and *transitions,* since splitting a scene gives you additional control over those effects.

■ Splitting your movie into more scenes does not adversely affect the file size, so you should feel free to create as many scenes as necessary to achieve the results you want.

■ When you are editing your movie, a scene is the smallest increment that you can modify independently. Any changes you make to a selected scene affect the entire scene. This is another good argument for using more rather than fewer scenes.

■ Scenes do not have to appear in the order in which they were shot. I've mentioned this before, but it is such an important point that I wanted to reinforce it. Sometimes it just isn't possible to film a story from beginning to end without some backtracking. Being able to adjust the sequence of your clips gives you a whole lot of flexibility.

By now you should have a much clearer understanding of the role of storyboards, timelines, and scenes in your movies. You have seen that each of these elements provides an essential part of your video editing toolbox. Storyboards give you an easy method of visualizing the sequence of the scenes, while timelines provide more clues to how long the various scenes will play. Scenes, of course, are the building blocks of your movies.

Next we'll have a look at one of the more interesting and fun areas of digital video editing: applying transitions and special effects.

CHAPTER 5

Transitions and Special Effects

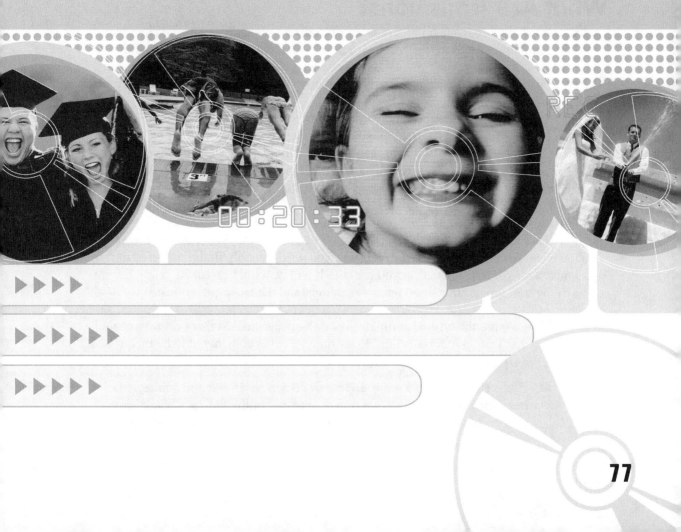

▶▶▶▶▶

▶▶▶▶▶▶▶

▶▶▶▶▶▶

Now that you've had the chance to learn the basics of storyboards, timelines, and scenes, it's time to move on to two of the more interesting and fun areas of digital video editing. In this chapter, we're going to have a look at *transitions* and *special effects*. These are the things that fall into the "how did they do that" category when you see them onscreen, but as you'll soon learn, they're very easy to apply in most digital video editors.

It certainly could be said that transitions and special effects are frills rather than necessities in the movie-making process. Still, I think that these frills can enable you to produce far more interesting and professional-looking results—especially once you understand how they work and how to best use them.

What Are Transitions?

Most movies are made up of a number of different scenes, of course. When you first add your scenes to the storyboard or timeline, there are no scene transitions, so there is an abrupt change on the screen when one scene ends and another begins. A transition eases the change from one scene to the next, so that the effect is much smoother.

To get a better idea about how transitions work, take a look at Figure 5-1. I've added a right-to-left push transition between the second and third scenes on the timeline. As the movie plays through the transition, scene two slides to the left as scene three slides in from the right side of the frame (you can see this in the viewing window in the upper-right area of the Pinnacle Studio window). In this case, it seems as though scene three is pushing scene two out of the frame—that's why this type of transition is called a *push*.

There are quite a few different types of transitions that you can use in your movies. As a beginning movie producer, you will probably find that there are far more types of transitions than you can use. It will be a little easier for you to decide on the best type of transition once you understand the basic categories:

- *Cuts* are the type of abrupt transition between scenes that occur when you don't add a *real* transition. Essentially, a cut is considered to be the lack of a transition.

- *Fades* slowly make a scene appear or disappear. If you add a fade at the beginning or the ending of a movie, one end of the fade is a black screen and the other is the adjacent video clip. When you add a fade between scenes, the screen goes dark between the two scenes.

FIGURE 5-1   This shows a scene transition where the scenes slide right to left.

■ *Dissolves* are similar to fades, except that the screen doesn't go completely dark between the two scenes.

■ *Wipes* replace the first scene with the second scene by slowly revealing the second scene. In a wipe, neither of the two scenes seems to move across the frame. Figure 5-2 shows an example of a wipe transition where the second scene is wiping up from the bottom of the screen.

■ *Slides* are quite similar to wipes, except that the second scene slides into position over the top of the first scene. In a slide, the first scene does not seem to move, but the second one does. Figure 5-3 shows a slide transition where the second scene is sliding up from the bottom to cover the first scene.

FIGURE 5-2 This is a wipe transition.

- *Pushes* are similar to slides, except that both scenes appear to move. That is, the second scene seems to be pushing the first scene out of the way. Figure 5-4 shows an example of a push transition.

FIGURE 5-3 This is a slide transition.

FIGURE 5-4   This is a push transition.

In addition to the standard transitions, you'll find many others that you can use. For example, Pinnacle Studio offers quite a few very fancy transitions, including some they call Alpha Magic and several Hollywood FX categories. In most cases, however, you'll be able to identify each of the very fancy transitions as belonging to one of the basic categories I mentioned earlier. In fact, so that you'll have a better feel for transitions, I encourage you to take a few minutes playing around with some of the fancy transitions to see if you can identify the basic category they belong in. You'll soon have a clearer idea about which types of transitions will fit in your movies, and you won't be so intimidated by all of the choices.

When you are thinking about which types of transitions will work the best in your movies, keep in mind that the overuse of transitions can be pretty distracting. Remember that transitions should be used where they can enhance your movie, rather than used as something you throw in simply because you can. In other words, don't make transitions so prominent that they are all anyone remembers after they've viewed your movie.

Applying Scene Transitions

Now that we've had a look at what transitions are, let's go ahead and add some to our movie. This is actually a very simple process and is similar to adding movie clips to the timeline.

Transition Considerations

Adding a transition to your movie may have effects that you had not considered. Depending on the purpose of your movie, these effects could be quite important. For example, if you are creating a movie that will be used in a very time-specific venue—such as a TV commercial—it is important to understand how adding the transition will affect the overall length of the movie. Similarly, if your movie depends on the recorded sound track, you must take into account how a transition affects this, too.

Here are some important things you need to know about the effects of adding transitions to your movies:

- Adding a transition actually shortens the length of your movie by the length of the transition itself. This may seem counterintuitive, but it is true. The reason for this is that when you add the transition, the second clip actually begins playing at the beginning of the transition, while the first clip continues to play until the end of the transition. By default, transitions last for two seconds, so if your movie started out as 60 seconds long, adding two transitions would shorten the movie to 56 seconds. And if you lengthen the transition, you will shorten your movie even more.

- Transitions also affect the audio track that was recorded with each movie clip. As a transition begins, the audio from the first clip is slowly faded down, so that it reaches a zero point at the end of the transition. Likewise, the audio from the second clip is slowly ramped up until it reaches full volume at the end of the transition. Therefore, as the transition is occurring, your audio track will actually be a mix of the audio tracks from the two clips. Needless to say, this could be a little confusing, especially if the transition is longer than normal or if the audio tracks are especially loud. A fade transition, however, does not mix the two audio tracks, because both the audio and video tracks go to a zero level in the middle of the fade transition.

- Transitions also are limited in length. This limit is determined by the shorter of the two adjacent clips and is one frame shorter than that clip. Generally speaking, however, this limit probably won't have much effect, since a transition that takes up an entire clip would probably not be very desirable anyway. Note, though, that if you add another transition at the other end of the clip, that transition also must use part of the clip, which further limits the length of the transition.

Adding Transitions

Adding the transitions to your movie is very similar to adding clips to the movie. The primary difference is that when you open the transitions album, you see a representation of the transition effect, rather than a scene from your movie.

Figure 5-5 shows the basics of adding a transition to the timeline. Let's briefly review the steps.

Click this tab to open
the transitions album.

Select the transition
from the album pages.

Choose the set of
transitions here.

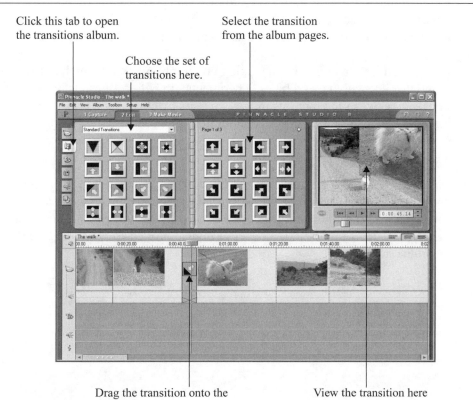

Drag the transition onto the
timeline between the two clips.

View the transition here
by playing the movie.

 FIGURE 5-5   Use the transitions album to add your transitions.

1. Open the transitions album by clicking the Show Transitions tab. You can also open this album using the Album | Transitions command.

2. Select the set of transitions you want to use from the drop-down list box on the top of the left-side page of the album.

3. Drag the transition you want to use onto the timeline and drop it between the two clips where you want the transition to appear.

4. Use the playback controls to view the transition in the viewing window. By the way, this is one place where you will really appreciate the fine control that is possible when you are using a Contour ShuttlePRO!

If you don't care for the transition that you have selected, you can replace it simply by dragging a different transition into the same place on the timeline. Or if you have tried out a number of different transitions and have decided that none of them seem to work, you can delete the selected transition by clicking on the trashcan icon.

Don't feel bad if you have trouble making up your mind about which transition to use. I find that it usually takes a bit of experimentation before I'm satisfied. I try to strike a balance between the motions in the two clips by choosing a transition that moves in the same general direction as that motion, if possible.

Adjusting Transitions

As I mentioned earlier, the default length of the transitions you add to your movies is two seconds. Depending on the effect you want to produce, this may be too short or too long. If so, you can adjust the length of the transition by dragging the beginning or the ending of the transition as shown here (the double-headed arrow is dragging the right side of the transition).

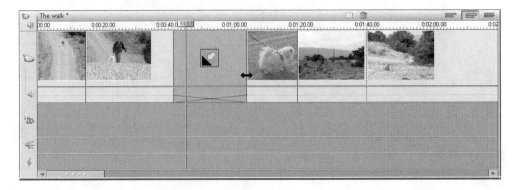

As you drag the ends of a transition to adjust it, keep the following important points in mind:

■ The viewing window will show the frame where the transition will begin or end (depending on whether you're dragging the beginning or the ending of the transition). This does not mean, however, that only one end of the transition will change, because transitions always extend their full length into both of the neighboring movie clips.

■ As you lengthen a transition, the overall length of your movie decreases by the same amount as the additional length of the transition.

■ If you make a transition too long, the shorter of the two clips may be completely taken over by the transition.

■ Very long, slow transitions can be used to simulate the passage of time.

■ Very short transitions can give your movie a very busy, frantic feel.

■ Since the audio tracks from the two clips cross fade during the transition, you may want to separate the audio tracks (as discussed in Chapter 6) from the video clips—especially during long transitions which include dialog—so that you'll have better control over the audio effects.

Feel free to adjust the length of your transitions as necessary. After all, you are the person who has the best idea about the results you want to achieve.

Rippling Transitions

There may be times when you want to apply the same transition between several of the scenes in your movie. This can be very effective if your movie is actually a slide show consisting of a series of still images and you want a consistent effect over that series of slides.

In applying the same transition to a series of clips, you would fine-tune the first transition to get just the transition length you want and then use exactly those settings for each clone of the transition. That way, you won't have to individually adjust the transitions after they've been placed between the clips.

To apply the same transition to a series of clips, follow these steps:

1. Add the desired transition between the first two clips where you want to use the transition.

2. Make any adjustments necessary to the length of the transition.

3. When you have finished your adjustments, click the second of the two clips so that *it* is selected rather than the transition.

4. Hold down the SHIFT key and then click the final clip that you want to include in the series of clips, using the same transition to select the whole series.

5. Right-click the selected clips and choose Ripple Transition from the pop-up menu, as shown here. This will add the same transition between each of the clips.

Delete	
Cut	Ctrl+X
Copy	Ctrl+C
Paste	Ctrl+V
Split clip	
Combine clips	
Find Scene in Album	
Ripple Transition	
Remove Transitions	
Go to Title/Menu Editor	
Clip Properties	
Set Disc Chapter	
Set Return to Menu	
Set Thumbnail	

It's probably best to limit your use of the Ripple Transition feature. It really is most effective for creating a slide show of still images rather for applying the same transition between a bunch of video clips. This is simply a matter of esthetics, of course, but I think that transitions work best when they're such a natural part of your movie that a person viewing it doesn't give them a second thought.

Editing Transitions

If you have one of the more complete digital video editing packages (such as Pinnacle Studio Deluxe), you may be able to take your creativity to even higher levels by modifying the transitions to suit your individual artistic leanings. Here, for example, I'm using *Hollywood FX Plus* to fine-tune a particular transition so it looks just the way I want it to:

If you have Hollywood FX Plus (or Pro) installed on your system (not the basic version of Hollywood FX), you can edit your transitions by following these steps:

1. Apply the transition to the timeline where you want it to appear in your movie.

2. With the transition selected, click the Video Toolbox button.

3. Click the Edit button that appears in the upper-center area of the video toolbox, opening the Hollywood FX editor window I just showed you. (If the Edit button does not appear in the video toolbox, you have the basic version of Hollywood FX installed and cannot edit your transitions.)

4. Make any adjustments you want to the transition.

5. Click the OK button to close the Transition editing window. Here, I've returned to Pinnacle Studio and am previewing the modified transition.

TIP
If your system crashes when you are editing transitions, you may want to use the Setup/Edit command and then turn off the selections for rendering as a background task and for using hardware acceleration. If doing this solves the problem, it may be possible to turn one (but not both) options back on—you'll need to experiment to see which one is the culprit.

Using Special Effects

The power that is packed into today's digital video editors is truly amazing. One place where this power is very evident is in their ability to apply special effects to video clips. At first glance, you may not think that the ability to adjust the color, brightness, contrast, focus, and so on is anything special, but when you realize that your PC adjusts these properties for 30 frames for each second of video, you get an idea of just how awesome a task this really is.

There are any number of reasons why you might want to use some special effects in your movies. Consider these possibilities:

- A scene may be too dark for easy viewing. Adjusting the brightness and contrast may make it possible to use your existing footage. This could be especially important in the case of old family movies that cannot be redone or for videos you might have made at a concert.

- You might want to turn a section of the movie into black-and-white or sepia tones to produce the effect of vintage films. This can be especially effective for flashback scenes in a movie that is otherwise in full color.

- A scene shot under poor lighting conditions (such as fluorescent lighting) might need some color adjustments for a more natural appearance.

- Using a blur or a mosaic effect can help disguise the identity of people in a scene. This might be useful if you are creating a video that you'll use commercially but for which you weren't able to get model releases from everyone who appears in the video.

You may find that only parts of a scene really needs some special effects applied. If so, simply split the scene into as many pieces as necessary so that you can make finer adjustments. You may also want to experiment with splitting a scene into a number of very short scenes and then applying incrementally greater or smaller adjustments to the subsequent scenes. This could be an effective way to gradually return a scene to full color, for example.

TIP	*If you want to make gradual adjustments to a particular property in a scene, begin by adjusting that property for the whole scene. Then split the scene where you want to begin varying the property and make a small further adjustment to the remaining portion of the original scene. Continue splitting and then making small adjustments until you've made the complete adjustment.*

Figure 5-6 shows the color adjustment pane of the video toolbox in Pinnacle Studio. Let's take a closer look at what each of these different adjustments can do.

Select the
color type here.   Hue
                   Saturation
                      Brightness
                         Contrast

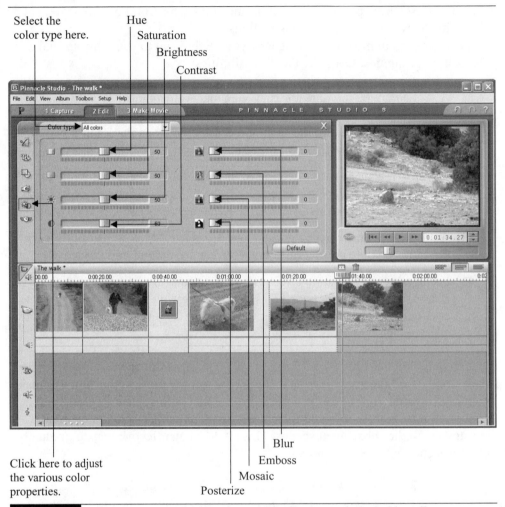

Click here to adjust   Blur
the various color   Emboss
properties.   Mosaic
                      Posterize

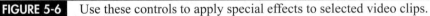

FIGURE 5-6   Use these controls to apply special effects to selected video clips.

- **Hue**   This control adjusts the color balance. Move the control left to add more red or to the right to add more green. You can use this adjustment to compensate for poor lighting.

- **Saturation**   Use this control to change the amount of color in the scene. Move the control to the left to reduce the color level or to the right to increase the color level. You can use this adjustment to gradually change a scene between monochrome and full color or to give scenes a "cartoony" effect by over-saturating the color level.

■ **Brightness**   This slider makes the scene darker as it is slid to the left or brighter as it is moved to the right. You could make a scene shot during daylight appear to be a night scene by reducing the brightness.

■ **Contrast**   This control adjusts the relative brightness between light and dark objects in the scene. It could be handy for making scenes shot on overcast days appear more like scenes shot during full sun or for reducing the excessive contrast that sometimes affects beach scenes.

■ **Blur**   You can use this control to make it appear as though the camera was out of focus. This could be especially effective in a filler scene where you want the viewer to listen to the narration rather than concentrating too much on the video elements. Here is an example of the blur control set to full:

■ **Emboss**   This control makes the objects in the scene appear as though they were embossed onto thick paper. Here, too, this special effect is probably most useful for transition scenes, since the viewer won't really be able to recognize very much (as you can see next).

■ **Mosaic**   You can use this control to produce a very blocky appearance, as shown here. This could be used to disguise people or places in a scene, giving the effect of a crime scene video or a secret agent movie.

■ **Posterize**   This final control increases the contrast while reducing the number of colors in the scene, as shown next. One use for this might be to simulate an alien planet if you were creating a Sci-Fi movie spoof.

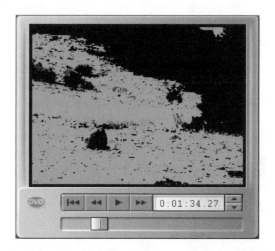

In addition to the individual slider controls, you can also choose a specific color type from the drop-down box at the top of the video toolbox. The selections you make here interact with the sliders, so you could produce a red planet effect by selecting single hue as the color type, then sliding both the hue and saturation controls completely to the right. Or you might simply choose the sepia option to make it look as though your movie was from the early days of film-making.

In most cases, you'll probably find that a gentle touch on the special effects controls will produce superior results. Of course, you may want to go wild and create a really bizarre-looking movie, but unless that's your goal, small adjustments should do the trick.

Transitions and special effects probably won't be a part of every movie that you make, but they do have their place. In this chapter you've learned how these items can affect your videos, and you've seen how to make good use of them.

Next we'll look at your audio options. You'll see that the audio track that was recorded along with your original video clips is really just a basic starting point. You can add narrations, background music, sound effects, and so on for a far more professional effect. You'll also learn how to make better use of the recorded audio track as well as how to blend the various sound sources together.

CHAPTER 6

Audio Tracks for Your Movies

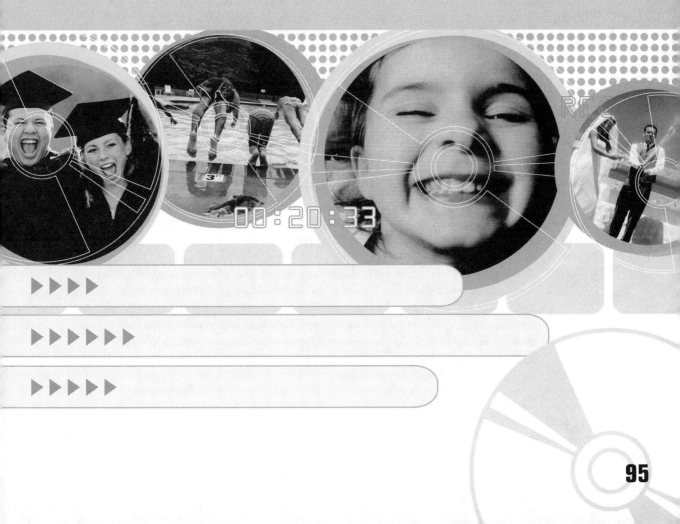

▶▶▶▶

▶▶▶▶▶▶

▶▶▶▶▶

In the early days of movie making, all movies were silent because it simply wasn't possible to record sounds along with the pictures. Soon, however, audiences wanted something more lifelike, and the "talkie" was born. Making movies hasn't been the same since!

These days people expect a *multimedia* experience. They want to hear some dialog, some background music, and maybe even some sound effects to make them feel as though they're watching a live video performance. As a digital video producer, you need to know how all of these pieces fit into your movies so that you can make the best use of them. In this chapter, we'll examine the tools and options that are available to help you make movies that not only look good, but sound good too.

As with all of the elements that make up a successful video production, it's important to consider your target audience as you plan the audio portions of your movies. This, however, is an area where you'll really have to use a lot of personal judgment. I can tell you how to manipulate the audio tracks, but I can't really tell you which types of music will be the best. Oh sure, I could tell you to use my favorites, but that would be like trying to dictate that all of the radio stations in the world could only play one genre. That would make for an awfully boring world, wouldn't it?

Audio Track Sources

You already know that your camcorder (or VCR) is the source of the video portion of your movies, but when it comes to audio you have a number of options. These options probably include several sources you may not have considered. Having so many options does make for a bit more complication, of course, but it also means that you can have a lot of fun adding audio to your movies.

> **NOTE** *Audio can be a real copyright nightmare, especially if you plan on any sort of commercial use for your videos. Basically, you should assume that any recorded audio source is covered by a copyright; any unauthorized commercial use is a very bad idea. Look for* royalty-free *audio licenses unless you want to go to the trouble of obtaining written permission to use audio recordings. You may not have to worry quite so much in your non-commercial videos, of course.*

Let's take a look at some of the sources you can use for your audio tracks. Figure 6-1 shows where each of these audio tracks appears on the Pinnacle Studio timeline.

Using Recorded Audio

By far the one audio track that you're most likely to use is the one that was recorded by your camcorder as you filmed your video. Using this recorded audio track takes no extra effort on your part, because it is automatically imported along with your video.

You may have noticed in Figure 6-1 that the recorded audio track appears just below the video track in the timeline. In addition, this audio track is shown with a dotted line connecting its icon to the video track icon (in the left-most column of the timeline). These are visual clues to something very important about the recorded audio track. Unlike any other audio track, this track is automatically *synchronized* with the video so that the sound will exactly match the video. For example, if someone is talking in the video, their lip movements and the sound of their voice will match up perfectly.

Even though the recorded audio track starts out perfectly synchronized with the video track, you are free to manipulate the two tracks independently of each other. For example, shown in Figure 6-2, I've adjusted the audio tracks that were associated with the first and second video clips so that they no longer are synchronized with the video.

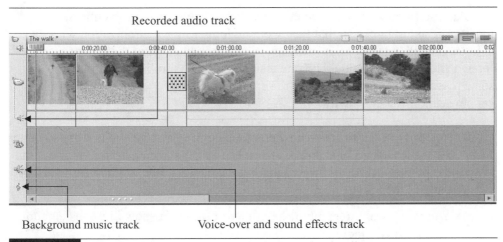

FIGURE 6-1   The three audio tracks also show up on the timeline.

FIGURE 6-2 After some manipulation, the recorded audio tracks are no longer synchronized with the video track.

Okay, so I know what you're thinking. Why would I want to deliberately modify the audio tracks so that they were no longer in synchronization with the video? After all, wouldn't this make my movie look like one of those poorly dubbed foreign films? Yes, this could be one result from such modification, but here are some reasons why you might want to separate the audio and video portions of certain scenes in your movies:

- You might want to cut some video footage from a scene while still including all of the audio. For example, if you had an off-screen narrator talking about the scenery while you were panning the camera, you might want to select parts of the video and other parts of the audio to use.

- You might want to combine a number of different shots of a speaker and still be able to include the entire speech.

■ In some cases, you may have footage with great audio effects that would fit perfectly elsewhere in the movie. An example of this might be the sound of a car accelerating and up-shifting that you might want to add to a scene at the beginning of a road trip movie (which is exactly what I'm doing in one of my movies).

■ It's even possible that you might want to shift the audio track forward or backward slightly to produce a humorous effect of the speakers' voices being out-of-sync with their mouths. I can imagine all sorts of interesting ways to make use of this effect.

■ You might want to separate the audio and video tracks if you are going to use slow-motion or fast-motion video effects, since the recorded audio won't be heard if it is synchronized to a video track that is being played at other than the normal speed.

I'll show you how you can separate the recorded audio track from the video track as well as how you can manipulate the audio tracks later in this chapter.

Ripping Audio CDs

In addition to the recorded audio track, you may want to spice up your movies by adding some background music. One source you may consider for this music is your favorite audio CD. This may be especially appealing in the case of a family movie where you might like to use a cut from an audio CD that your family members somehow associate with a particular event.

Adding music from an audio CD is a fairly simple process. Figure 6-3 shows the tools that you use in Pinnacle Studio.

Let's take a look at the process in detail:

1. Begin by selecting the video clips that you want to include with the new background music track.

2. Click the audio tool box icon above the upper-left corner of the timeline.

3. In the audio toolbox, click the "Add background music from an audio CD" icon to open the set of tools shown in Figure 6-3.

4. If you have not already done so, insert your audio CD. If Studio does not recognize the CD automatically, you may have to type the name of the CD.

Click here to add background music from an audio CD.

Choose the CD and track here.

Click here to add the song to the timeline.

Adjust the song's duration here.

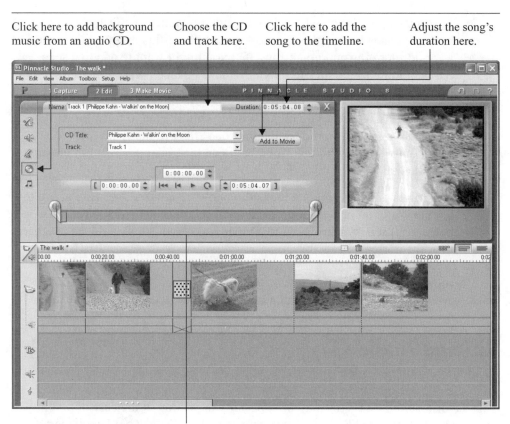

Drag these markers if you want to trim the song.

FIGURE 6-3 This is how you add music from an audio CD to your video.

5. Select the track that you wish to add from the drop-down box just below the name of the CD.

6. If necessary, use the trim markers or the duration control to set the length of the audio clip you want to add.

7. Click the Add to Movie button to add the track to the timeline. Although this will seem to add the music to the background track on the timeline, it actually is only putting a placeholder on the timeline for now.

8. Click the Play button below the viewer window to begin capturing the audio track, as shown here:

9. Once the track has been completely captured, you can continue editing your video, as shown here. As you play the movie now, the background music will play along with the video.

As you have seen in this example, you can add an audio track that spans as much of the video as you like. That is, you could add background music to a single scene, to several consecutive scenes, or even to the entire movie.

Using Pro Soundtracks

I have, of course, warned you about the copyright implications of using music from commercial CDs in any sort of commercial use video you might produce. Frankly, however, obtaining the necessary legal permission to use someone's music in your video can be both difficult and costly. Still, this should not prevent you from having background music in your video. The solution is simply to use royalty-free music that you can use without any special permission (this is also sometimes called *stock* music).

Pinnacle Studio includes royalty-free music in the form of something called *SmartSound*. These instrumental pieces can be fit to the length of virtually any video clip, and you can add them to any of your movies—even commercial productions—without worrying about breaking any copyright laws.

Adding a SmartSound track to your movie is actually quite similar to adding music from an audio CD. The primary difference is that you get to choose the style, then the song, and finally a version of the song. As is shown in Figure 6-4,

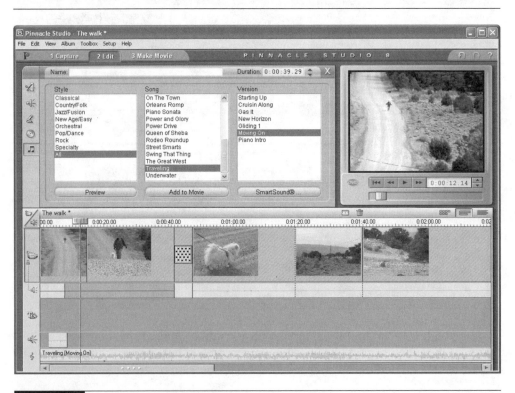

FIGURE 6-4 Choose background music for your movie.

you begin by making a choice on the "Add background music automatically" tab of the audio toolbox.

> **TIP** *When you first add background music tracks to your video, you may be surprised at how loud it seems. But as you'll soon see, it is very easy to adjust the volume of each of the tracks to get the results you want.*

Before you add a SmartSound track to your movie, make sure that you select exactly those video clips you want the music track to span. If you do, the program will automatically generate a sound track of the correct length. I find that if I make a change to the length of a video clip or if I rearrange the video clips, it is easiest to just delete the background music track and then create a new copy of the correct length. In terms of efficiency, you should complete your video layout and editing before you begin adding music to your productions.

Using Recorded Sound Effects

You might want to add sound effects to your audio tracks. These are short clips that cover the whole range of non-musical audio samples. For example, you could add a section of laughter or applause to a video of someone giving a talk. You might add the sound of crickets to set the mood for a movie about your recent camping trip, or of seagulls to a video of your trip to the ocean.

You can, of course, record sound effects yourself. Fortunately, though, you probably won't have to do so because there is a wealth of sound effects you can simply plug into your movies. As shown in Figure 6-5, I've opened the sound effects album in Pinnacle Studio and am considering adding some animal sounds.

There is one potentially confusing element of adding sound effects to your movies: in order to add sound effects, you must first close the audio toolbox and then open the sound effects album. Aside from this, you'll find that procedures for sound effects are quite straightforward. In most cases, though, you'll probably want to trim the sound effect after you've added it to your movie, since the sound effects often include a couple of repetitions (which are usually slight variations on a basic theme).

> **TIP** *Be sure to place the timeline marker (also called the* scrubber*) exactly where you want the sound effect to play before you add that sound effect. Use the viewing window to locate the precise frame where the sound effect should begin.*

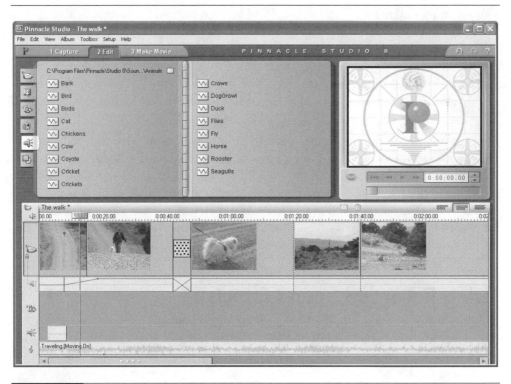

FIGURE 6-5   You can choose from a broad selection of recorded sound effects.

Adding a Narration Track

The final type of sound track that you might want to add to your movies is the narration (or voice-over) track. This track actually shares a track with any sound effects you may add, so it may take some careful coordination to combine both sound effects and narration in your movie to get exactly the effect you want.

Narration is generally added to a movie so that an off-stage announcer can provide details about what the viewers are seeing on the screen. You might use this for a number of interesting purposes:

- Narration can be very effective in giving a historical background during portions of a movie where you present a slide show of still images.

- A voice-over track is an effective tool for setting up a scene. For example, you might use this feature to explain that the movie is showing the oldest Monterey Cypress in the Big Sur area.

■ Narration tracks can also be used to add the standalone recording of a speaker who was too far away for your camcorder's microphone to pick up properly.

■ Finally, a voice-over might be the only effective way to combine a recording of a long dead relative with some old silent family movies that you've converted to digital video.

To add a voice-over track, you use the recording tools shown in Figure 6-6. Let's take a look at the specifics of adding narration to your movie:

1. Position the timeline scrubber at the frame where you want to begin recording the narration.

Click here to open the voice-over tools.   Watch for the recording countdown here.   Drag this slider to adjust the recording volume.

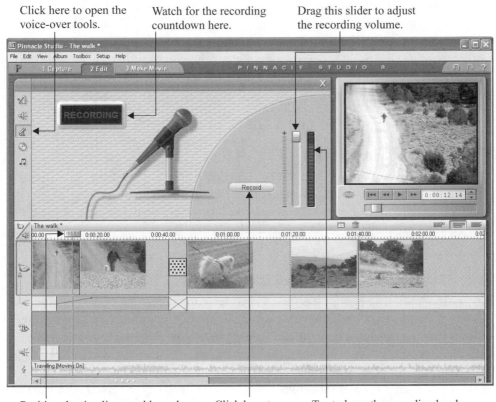

Position the timeline scrubber where you want the narration to begin.   Click here to begin recording.   Try to keep the recording level in the yellow range of this scale.

FIGURE 6-6   These tools enable you to add narration to your videos.

2. Click the audio toolbox icon to open the audio toolbox (if it is not already open).

3. Click the microphone icon on the left side of the audio toolbox to display the recording tools shown in Figure 6-6.

4. When you are ready to begin, click the Record button. This will start a three-second countdown (you'll see the countdown indicators in the upper-left area of the audio toolbox).

5. Once the countdown concludes, the recording light will come on. Begin talking when you see the red recording light.

6. If necessary, adjust the volume level slider to move the recording level indicator into the yellow band.

7. Click the Stop button when you are finished with your narration.

You may want to write a script for your narration and practice it before you begin recording. It's usually easier to get the results you want if you do a dry run or two beforehand. Also, don't be afraid to delete the track and record it again if you aren't happy with the results.

Manipulating Audio Tracks

You have already seen that video tracks sometimes need some editing in order to produce the results you want. The same thing applies to the audio tracks. If you want your movie to be the best it can be, you'll probably want to spend some time doing careful adjustments to the audio tracks.

Adjusting the Volume Levels

The most common as well as the most complicated adjustments you're likely to make to the audio tracks will be volume adjustments. Quite frankly, balancing three different audio tracks can be a bit challenging until you've become comfortable with the process.

A major part of the confusion regarding adjusting volume levels comes from the fact that there isn't just one simple adjustment. Each of the three audio tracks has independent adjustments, and you have several options for each of them (see Figure 6-7).

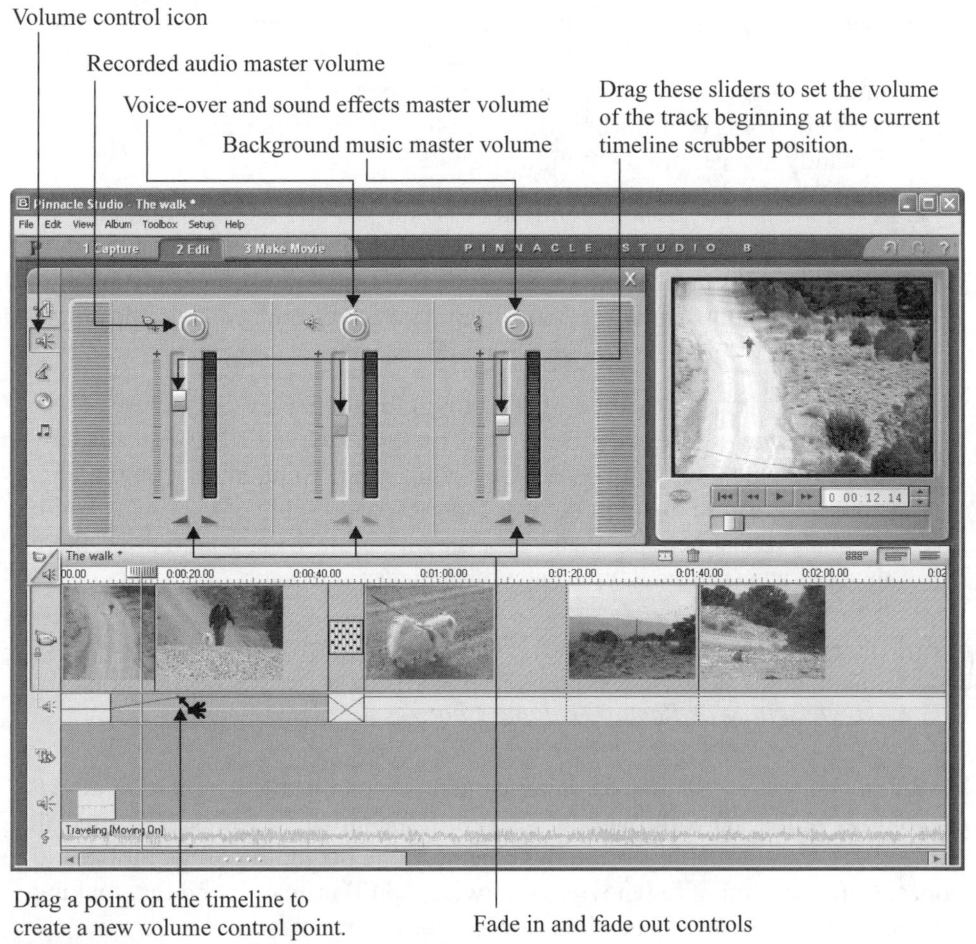

Volume control icon

Recorded audio master volume

Voice-over and sound effects master volume

Background music master volume

Drag these sliders to set the volume of the track beginning at the current timeline scrubber position.

Drag a point on the timeline to create a new volume control point.

Fade in and fade out controls

FIGURE 6-7   Adjusting the volume levels can be a little confusing.

To give you a better idea how all of this works, let's take a look at each of the volume adjustment possibilities:

- Each audio track has a *master volume* control that raises or lowers the level of the whole track as you turn the knob. This master control is a good starting point, since you can use it to equalize the volume levels so that they are roughly the same. This is a good way to lower the level of the background music track, since it is often too loud at first.

■ The *sliders* raise or lower the level of the track beginning at the current position of the timeline scrubber. Unlike the master volume controls, the sliders do not change the level before the position of the timeline scrubber. This control is useful when you want to take the volume up or down instantly at a certain point in the movie.

■ The *fade in* and *fade out* controls cause the volume levels to ramp up or ramp down (respectively) from the current position of the timeline scrubber. This makes for a more gradual volume adjustment than using the slider and is handy for preventing the audio from suddenly blasting out at full volume or in suddenly being cut off.

■ The most versatile method of adjusting the volume levels is to drag the *blue volume line* directly in the track on the timeline. When you click and drag the blue line, you create a new adjustment point on the volume line, and this creates a bend in the line (as shown in Figure 6-7). You can add as many adjustment points as necessary, thus allowing you to bring the level of an audio track up or down whenever you want.

> **TIP** *To remove volume level adjustments you no longer want, right-click them on the timeline and choose Delete Volume Setting to remove a single setting or Remove Volume Changes to reset the entire track.*

When you make a volume level adjustment, that adjustment generally affects the volume for that track throughout the rest of the movie. As a result, it's usually the best idea to start making your adjustments at the beginning of the movie and work towards the end of the movie. Otherwise, you'll probably find that making volume adjustments simply becomes far too confusing.

Splitting Recorded Audio Tracks from the Video Track

Earlier in this chapter I mentioned that even though the recorded audio track is automatically synchronized with the video track, there might be times when you would want to use the two tracks independently of each other.

There are a couple of ways to manipulate the recorded audio track separately from the video track. If you want the two tracks to remain synchronized but don't want them to be the same length, you can trim the recorded audio track and the video track separately. To do so, you lock the track that you don't want to change by clicking the track's icon in the timeline. Here I've locked the video track and have dragged the beginning of the recorded audio track, which is associated with the second scene to the left. As a result, the audio portion of the second scene will begin playing while the first scene video is still playing.

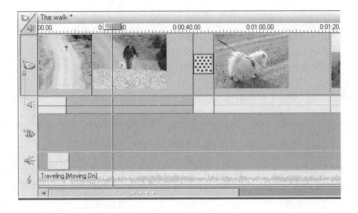

If, on the other hand, you want to use a recorded audio track and don't want it to be synchronized with the video, you can move that recorded audio clip to one of the other audio tracks. Here, I've removed the audio track that was associated with the second video scene and placed it a bit further along on the voice-over/sound effects track:

Since I already had a short piece of recorded narration earlier on that track, it was necessary to place the audio clip a bit past that narration. In this case, the recorded narration is to the left of the timeline scrubber on the voice-over/sound effects track, and the audio clip I moved from the recorded audio track is to the right of the timeline scrubber.

There are a few things to remember when you want to use this technique for manipulating the recorded audio track:

- Always lock the video track so that you can work with the audio track without affecting the video track.

- To remove the audio track so that you can paste it elsewhere on another audio track, right-click the portion of the audio track you want and choose Cut from the pop-up menu.

- To leave the audio track in place but still be able to use it elsewhere, right-click it and choose Copy from the pop-up menu.

- To place a copy of the audio track you've just cut or copied onto one of the other audio tracks, right-click a blank spot on the destination track at the location where you want to place the audio clip, then choose Paste from the pop-up menu. You can continue to paste the same audio clip into different locations as long as you do so before you copy anything else to the Windows Clipboard.

- If you no longer need the video clip that was associated with the audio clip, unlock the video track and then delete the video clip.

- When you are finished editing the audio, remember to unlock the video track—otherwise, you won't be able to make any changes to that track.

- You can lock any of the tracks to prevent changes to the track. You may want to do so once you have finished working on a track to protect your work.

Combining Multiple Audio Tracks

Each audio track is limited to holding just one audio clip at any point on the timeline, but all three of the audio tracks are actually able to hold any type of audio clip. You can use this framework to your advantage if you need to have multiple audio tracks of the same type.

You might, for example, need to combine two sound effects, such as a gunshot and the sound of a car racing away. As long as you can find blank spaces on two of the audio tracks, you can include both sound effects simply by using the cut and paste method I mentioned earlier. Remember that you can split a track using the razor blade tool and thereby make room on an audio track for your sound effects. For example, here I've cut out a section of the background music track below the third scene so that I'll be able to add a sound effect of a dog barking.

The audio portions of your movies really add a lot to the whole movie viewing experience. As you've learned in this chapter, having several distinct audio tracks enables you to create exactly the right blend of synchronized audio, narration, sound effects, and background music for your movie. If you've been following along and experimenting with your own videos, you've probably reached a point where you're really having fun making movies on your PC.

Next, we're going to turn to the process of adding titles to your movie. Titles can give a very professional touch, and they can also provide your viewers with important information about your movie.

CHAPTER 7

00:20:33

Titles for Your Movies

00:20:33

▶▶▶▶

▶▶▶▶▶▶▶

▶▶▶▶▶

Next, we'll turn our attention to titles for your movies. Good-looking titles can really add a professional touch and make a movie far more memorable. Who, for example, doesn't remember the opening title scene of the first *Star Wars* movie? The text flowing off into the distance creates an effect that anyone who saw the movie would instantly recognize, even if they didn't remember much else about the movie.

The subject of this chapter actually serves many different purposes. Even though we use the generic term "titles," what we're really talking about is any text that you might want to display in your movies. This could include actual titles, but it's likely to also encompass credits, captions, and even scrolling text that describes a particular scene or event. The idea here is to keep an open mind—just because we're calling them "titles" doesn't mean you can't use them for whatever purpose suits your needs.

In a digital video production, titles and menus are actually fairly closely related. Even so, I'm going to leave the subject of menus for the next chapter. In this chapter, we'll concentrate more on the cinematic use of titles. This way, you can concentrate on the titles themselves and not get confused by having the two subjects blended together.

Title Basics

So just what are titles, anyway? The simple answer might be text that you see on the screen—but that's really a bit too simplistic. By that definition, any signs you might happen to film would be titles, and clearly that's not what we're talking about. Rather, let's define titles as being "any text that you add to your movie during the editing process before you burn your disc." This is a broad enough definition to fit opening titles, closing credits, and any other text you might add somewhere in the middle. And, since we'll use the same techniques for each of these, you won't have to learn anything new when you move from one of them to another.

Understanding Title Types

Titles fit into two general categories, and each of them has their own unique characteristics. To a large extent there is little difference in the techniques you'll use for either of them, but it's important that you understand the basics before you rush off and begin creating titles.

Full-Screen Titles

The first type of title is the *full-screen* title. As can be seen here, a full-screen title takes over the entire screen in your movie and does not show your movie behind the title (although you can display a still image, such as a captured frame from your movie, if you like).

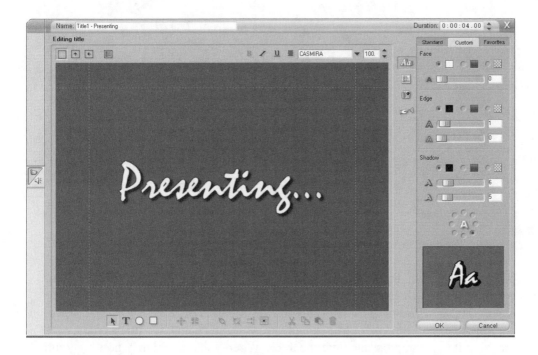

Full-screen titles sit on the main video track of the timeline, so any time that you use to display a full-screen title adds to the length of the movie.

Title Overlays

The second type of title is the *title overlay*, as shown next. This type of title has a transparent background so that it appears on top of—overlaying—a scene from your movie.

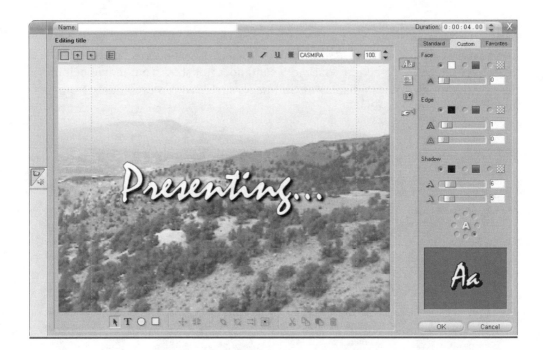

Title overlays are placed on the title track of the timeline rather than on the main video track. As a result, title overlays do not have any effect on the length of the movie. Also, the title overlays a moving image rather than a simple colored background or still image, so you probably need to be more concerned about readability than you might with a full-screen title.

Regardless of the type of title you create, you can change the title's type by dragging it to the appropriate timeline track. When you do, you may notice certain changes (such as the loss of a still image that you had placed behind a full-screen title). These changes will reflect the different nature of the two title types, of course.

Title Safe Areas

Most of the DVD movies that you create will likely be intended for playback on standard television sets. This factor has an important implication for any titles that you might want to add, because TV sets generally do not have quite the same characteristics as a computer monitor. In particular, any of the four edges of the screen may be cut off when your movie is displayed on a standard TV. In Figure 7-1, you can see that the title editor displays dotted lines that represent the safe area for titles.

These lines indicate the safe areas for adding titles.

These parts of the title are outside the safe area and might not display on a TV set.

FIGURE 7-1 Keep your titles inside the safe area lines to insure that they will be visible on TV sets.

I've also purposely created a title that extends beyond the safe area. If you added this title to your movie and then played it back on your PC, the whole title would appear and you wouldn't have a clue that there might be a problem. Unfortunately, if you then sent a copy of your movie to your grandmother to play on her set-top DVD player, it's very possible that she wouldn't see the two ends of the title that extend outside of the safe area. Although the whole title might show on some newer, high-quality TV sets, keep in mind that the safe area is designed to accommodate the vast majority of sets.

TVs and Title Readability

In addition to cutting off the edges of the film, TVs have a number of other bad habits that you need to be aware of when adding titles to your movies. Although it might not seem obvious, your computer's monitor has much higher screen resolution than you'll find on almost any TV. In fact, your monitor probably has higher resolution than even one of those fancy new high-definition TVs, which can cost thousands of dollars. Remember, screen resolution and screen size are two very different measurements.

Resolution is the number of individual dots that can be displayed horizontally and vertically. Higher resolution translates into finer detail being visible. Screen size is just the diagonal measurement of the screen without regard to how clearly the images are displayed.

In addition to having far poorer resolution than a PC monitor, TVs also suffer from the way images are displayed on a TV screen. Most TVs use an *interlaced* display. This means that as the image is drawn on the screen, only one half of the horizontal lines are drawn in a single pass. It takes two passes for the entire screen to be drawn. To make matters worse, it's every other line that is drawn in a single pass. As a result, the edges of any moving object tend to take on a jagged appearance on an interlaced display.

So how does this all relate to creating titles for your movies? The poor quality of most TV displays makes it very difficult to read small text—especially if that text is moving. To see this for yourself, take a look at the closing credits the next time you watch a Hollywood film on TV. Sure, you can easily read the names of the stars and the producer, but I'll bet you'll find that it's almost impossible to read the names of people like the caterers, the grips, and the Foley artists (since they tend to get much smaller print for their credits).

The bottom line is that you need to take the following precautions when using titles:

- Make titles large enough to read easily.

- Make titles fairly short so that they can be quickly understood.

- Use titles that are either still or moving very slowly, so they aren't too blurred.

- Place titles inside the title safe area so the edges of them aren't cut off.

- Use colors for titles that contrast well with the background.

If you don't follow these guidelines, the chances are good that your viewers will have a difficult time reading the titles when they watch your movie on a TV set. If you're going to all the work of creating titles, you'll want to make them readable!

Creating Titles

Now that you understand the basics of titles, let's go through the process of creating some of them. There are actually a number of different approaches you might take, but it seems to me that just digging in and having at it is probably the best way to see what is possible.

Even though full-screen titles and title overlays have some different characteristics, you can create either of them using the steps I'll show you. Where there are important differences, I'll make sure to point them out.

Selecting a Title Style

The first step in creating a title is to select the style for the title. You can do this by choosing an existing style from the title album or by starting from scratch and building the title yourself. In reality, both methods can produce identical results—the title album primarily serves as a starting point for people who need a little more help getting started. For example, in Figure 7-2, I've selected one of the sample titles from the title album and added it as a full-screen title at the beginning of my movie.

FIGURE 7-2 You can use one of the examples for the title album as the basis for your title.

Editing the Title

Although the title album is handy, you'll probably want more control over the contents and appearance of your titles. For that, you'll want to use the *title editor*. You can open the title editor in several different ways. The easiest two are to double-click a title on the timeline or to right-click either the main video track or the title track and choose Go to Title/Menu Editor from the pop-up menu.

Whichever method you use to open it, the title editor will look similar to Figure 7-3. Keep in mind that if you are working on a title overlay, you will see your movie behind the title.

As you work with different types of objects in your titles, you will find that various tools will be available at certain times but not at others. For example, the "move, scale, and rotate" control enables you to adjust the properties of a selected object such as a text box, but this control is only functional when you have selected

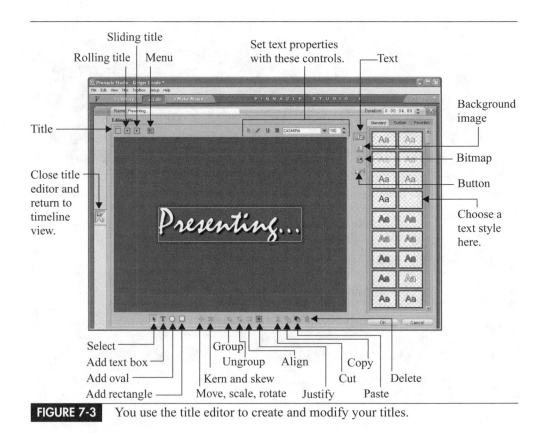

FIGURE 7-3 You use the title editor to create and modify your titles.

an object in your title. Likewise, the text manipulation tools are only available when you have selected a text box.

Editing the title should be pretty easy for you—especially if you've used a word processor or something like Paint. Basically, you'll follow these steps to add a text box and create a title:

1. Choose the style for the title from the examples along the right side of the title editor. Be sure to scroll down so you can see the entire list.

2. Click the text tool and draw a text box. Be sure to stay within the title safe area, as shown by the dotted line.

3. Enter the title text into the text box.

4. If necessary, adjust the text properties to get just the look you want.

5. When you are finished editing your title, click the video toolbox icon along the left edge of the title editor to return to the timeline view. Doing this will place the title on the timeline at the point where you right-clicked the timeline to open the title editor.

Be sure to check your spelling carefully. The title editor does not have a spell check, and a typo in one of your titles will certainly be very embarrassing!

Adding Background Images

As you'll recall from earlier in this chapter, one of the primary ways that full-screen titles differ from title overlays is that your movie does not show through a full-screen title. This doesn't mean, however, that your full-screen titles have to be boring. You can add color or even an image behind your full-screen titles.

Figure 7-4 shows how the title editor appears when you click the icon to add a background image to your title (this is the icon with a cactus on it near the upper-right corner of the viewing window). The background image album opens, so you can add an image behind your title.

You aren't limited to the images that are included in the backgrounds album—just click the pictures icon just below the backgrounds icon to select any digital image that is on your PC. For example, in Figure 7-5, I've used the Toolbox/Grab Video Frame command to save one of the frames from my movie as a bitmap image. I then added that image as the background for my title. Remember, though, that since this is a full-screen title, the image behind the title will be a still image rather than a video clip.

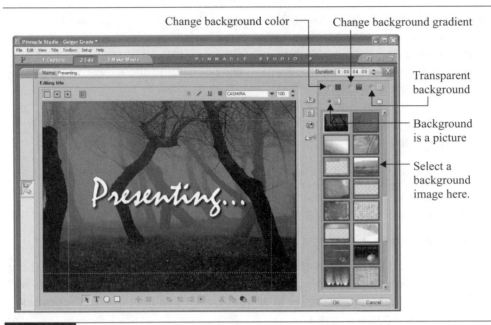

FIGURE 7-4 Use these options to add a background image to your title.

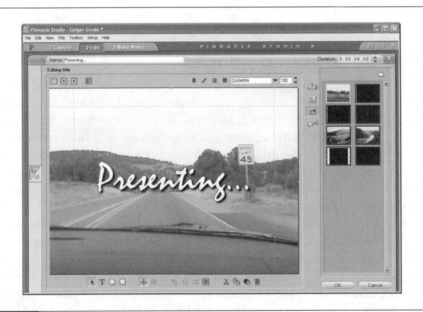

FIGURE 7-5 Here, I'm using a captured frame as the background for a title.

When you add a digital image to your title you can scale that image as necessary to achieve whatever effect you want. If you have already added a text box to your title, you may have to select the text box and then use the cut and paste buttons to place the text box in front of the imported image. The title editor automatically places new objects you add to the title in front of any existing objects, so you will want to plan carefully so that you don't accidentally hide an object behind another object.

Adjusting the Text Effects

Making the text in your titles readable always involves a certain amount of compromise. This is especially true when you use a title overlay, because the background can change as the scene plays underneath the title. In this section, we'll look at some text effects you can use both to make your titles easier to understand and also to add a little pizzazz to the text.

Figure 7-6 shows the tools that you use to adjust the text effects. In this case, I've used a full-screen title on a plain background so that you can more easily see

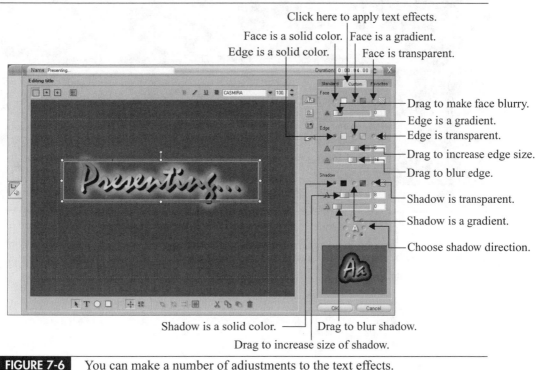

Click here to apply text effects.
Face is a solid color. │Face is a gradient.
Edge is a solid color. │ Face is transparent.

Drag to make face blurry.
Edge is a gradient.
Edge is transparent.
Drag to increase edge size.
Drag to blur edge.
Shadow is transparent.
Shadow is a gradient.
Choose shadow direction.

Shadow is a solid color. ──── │ Drag to blur shadow.
Drag to increase size of shadow.

FIGURE 7-6   You can make a number of adjustments to the text effects.

the text effects. You adjust the properties using the option buttons and the sliders. Dragging any of the sliders to the right side increases its value. So, for example, to increase the edge blur, you drag the lower slider in the edge section towards the right until the effect is just what you want. Be sure to click the text to select it before making your adjustments—otherwise those adjustments will apply only to new text boxes you add later rather than to the existing one.

As the figure shows, you can adjust the face, the edge, and the shadow of the text independently of each other. You can also combine several different effects to get just the look you want. Here, for example, I've increased the edge size to the maximum and have set all of the other controls to their minimum setting:

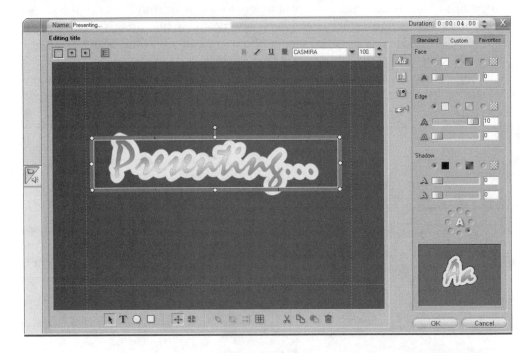

In contrast, here I've set the shadow to the maximum size with partial blur and set all of the other controls to a minimum:

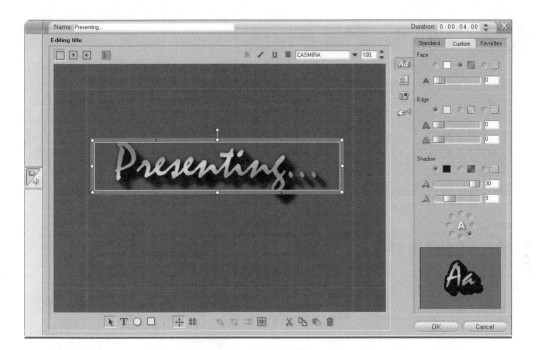

In practice, you'll probably find that some combination of the different adjustments produces the most pleasing effect.

TIP *Setting the edge to a large, blurred size can improve the readability of a title overlay, since doing this creates a zone of controlled brightness and color around the text.*

You may have noticed that the face section lacks the size slider that is included for both the edge and the shadow. The reason for this is simple—you adjust the size of the face by changing the font size using the box just above the right corner of the editing window. This is also where you can select the typeface and the other standard text attributes (as you learned earlier in this chapter).

Adding Title Special Effects

So far, you've learned how to create very nice looking titles that will serve many needs. But the titles you've created up until now were rather plain. That is, they just suddenly appear at a particular frame in your movie and then just sit there until they disappear just as quickly. They haven't exactly been the memorable *Star Wars* titles, have they?

In this section, we'll look at some techniques that you can use to spice up your titles a little. As with any special effects, though, remember that subtlety generally produces the best results. Don't try to hit your viewers over the head with fancy stuff. The best special effects usually are the ones you don't realize are happening, because they seem so real that you hardly notice.

Moving Titles

You can create titles that roll up from the bottom of the screen towards the top or that slide from the right side of the screen to the left (similar to a news ticker). Figure 7-7 shows a rolling title as it is being played in the viewing window—the title scrolls up as it is slowly displayed.

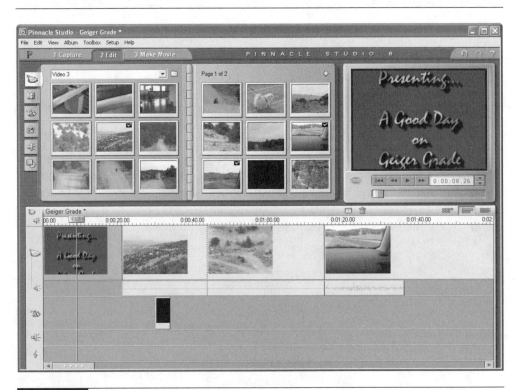

FIGURE 7-7   This title rolls up the screen as it plays.

Creating a moving title is quite simple. The primary difference between a moving title and a still one is that you must select the correct title type button (above the upper-left corner of the title editor window, as shown in Figure 7-3). You choose the roll button to create a title that rolls up from the bottom and the crawl button to create one which slides right to left.

You do need to keep certain things in mind when you are creating moving titles. These include the following:

■ For rolling titles, add as many lines to the title text box as necessary, but make sure that the text box does not extend beyond the title safe area lines at the sides of the screen.

■ Crawling titles can be hard to read unless you use a simple typeface.

■ Although this may seem counterintuitive, crawling titles do not need to be created on a single line in a text box. The title editor will automatically make the text into a single line when the title is played.

■ It is usually necessary to increase the duration of moving titles from the default of four seconds. Titles that move too fast won't be readable— especially on a standard quality TV screen. I generally figure out how long the title sequence should be by playing the title back and reading it as it goes along. If the speed seems comfortable, I then add at least 50 percent to the duration of the title sequence—after all, I already know what the title will say but my viewers will be reading it for the first time.

■ You may want to use the Toolbox/Grab Video Frame command to capture the final frame of your moving title and then add that frame as a still image scene lasting several seconds between the end of the title and the next scene. Otherwise, there will be a sudden jump from the end of the title directly into the following scene (unless you add a title fade, as I'll discuss next). By adding a still image at the end of the title sequence, you can make the title seem to pause for a short while as the audience finishes reading. (This applies to full-screen moving titles, of course.)

You'll almost certainly have to do a bunch of experimentation to get your moving titles into the condition you want. Always keep readability at the top of your list of objectives, since a title that no one can read is useless.

Title Fades

Because full-screen titles are a part of the main video track, they do offer one thing that is not possible with title overlays. You can apply any of the transition effects you learned about in Chapter 5 between full-screen titles and the adjacent scenes. In Figure 7-8, for example, I've added a dissolve between the full-screen title and the first scene of the movie. As a result, the end of the title slowly fades away as the first scene comes into view.

If you decide to add a transition other than a simple fade or a dissolve, you'll probably want to avoid using a moving title. Otherwise, it's likely that the end of the title sequence will be playing while it is being pushed off the screen or covered up by the next scene, and this may make the title impossible to read.

Creating Fancy Title Effects with Title Overlays

It's also possible to apply some pretty fancy transition effects to title overlays, but unlike the transitions between full-screen titles and other scenes, these transitions affect only the title overlay. That is, the transitions you apply to title overlays don't have any affect on the main video track.

FIGURE 7-8   Here the title is fading as the first scene appears.

Adding a transition to a title overlay involves a slightly different process than applying transitions to the main video track. As you'll recall from earlier discussions about the timeline and about transitions, there can be no gaps between the scenes on the main video track. This is not the case on the titles track where title overlays exist. You can place title overlays anywhere you like on the titles track. Once you have a title overlay on the titles track, you can add opening or closing transitions simply by dragging those transitions onto the titles track so that they are adjacent to the title overlay.

As shown in Figure 7-9, I've added a standard slide right transition at the beginning of my title overlay. This causes the title to slide in from the left of the screen and then stop once it is in place. Next I added the BAS-Scissors transition to the end of the title overlay so that it appears as though a scissors is cutting through the title at the end (see the view window to get an idea how this looks).

You can actually use any of the transition effects on title overlays, but because they will only affect the title overlay itself, the effect may be slightly different from when the same transition is used between scenes on the main video track.

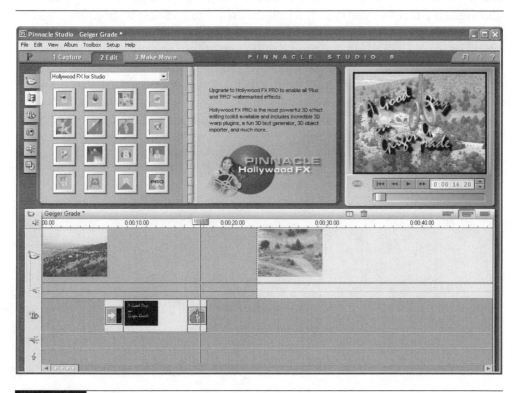

FIGURE 7-9   Here is an example of using a really fancy title effect.

It's possible to create some pretty awesome title effects using some of the fancier transitions with your title overlays. Rather than taking the time to discuss them in detail here, I'll simply remind you that Chapter 5 covers the subject of transitions and shows you how to edit them to suit your needs. Since these same effects also apply to transitions you apply to title overlays, I'm sure with just a bit of work you can end up with titles as fancy as anything you'll see anywhere.

> **NOTE** *Really fancy title effects take a lot of computing power to render. Even very fast computers can take a few moments to finish creating some of the really complex title overlay transitions. If you see a blue bar on the time scale above the timeline over your transition, it means that the transition is being rendered. Wait for the bar to disappear if you want to see the title transition at full quality.*

In this chapter I introduced you to a number of very important topics regarding adding titles to your movies. Because titles are so useful in helping to tell the story, you'll probably find that the little bit of extra work that goes into creating good looking and effective titles is well worth the effort. In addition to discussing basics like title safe areas and readability of titles on a TV, we also looked at some more advanced ideas, such as moving titles and applying fancy transitions to title overlays. You'll probably want to spend more time just trying out all of the possibilities.

Next, we're going to dig into creating menus for your movies. As you'll see, both DVDs and VCDs can have menus, but there are some important differences in the ways they function. You'll find that adding menus is an important element in making your movies viewer friendly.

CHAPTER 8

Menus for Your Movies

▶▶▶▶

▶▶▶▶▶▶▶

▶▶▶▶▶▶

If there is one area in which digital video offers the clearest change compared to the old analog video distribution methods, it is probably in the ability to have menus that enable the viewer to instantly access different parts of the movie. This capability makes it far more convenient for viewers to access a slide show, different product videos, specific lessons, or whatever you might be including on your disc— all using their set-top DVD player (or their PC). As a result, you are free to be far more creative with your content, since it's not necessary to produce one long video that tells a single story in a strictly linear fashion. By adding menus, you are opening up a whole new world of possibilities that were completely impractical (to say the least) with analog media such as videotapes.

Once you begin adding menus to your movies, you need to think about the movie timeline in a whole new way. In this chapter, I'll make sure that you gain a clear understanding of exactly how menus and their related links function. With this new understanding, you'll soon be able to take a completely different view of how to lay out your movies. Sure, you'll still be able to create features that simply run from start to finish, but you'll also be able to add some extras that simply wouldn't fit in the normal flow of the movie. For example, you can easily add an outtakes chapter or a special greeting from one of the actors. You could add a special section with information about how to join your organization, or you could have separate chapters telling about specific activities of your club members.

If you aren't excited about the possibilities that menus offer you yet, hang on. By the end of this chapter you will be. Let's dig in and see why I'm so sure about this.

Understanding How DVD Menus Work

So just how do menus produce all of this magic? How can they make it possible for you to add so many different features to your movies? To begin answering these questions, let's consider what a DVD really is.

What Are DVDs?

We already know that a DVD can contain a whole movie, complete with video, sound, titles, and so on. We also know that DVDs look an awful lot like audio CDs, that they can be played in set-top DVD players, and that they can be created on a PC that has a DVD burner and the proper software.

What you may not have considered, though, is that DVDs contain data in a digital format. In other words, DVDs store data in much the same manner as the hard drive does on your PC. As a result, it is very easy to access that data randomly. You don't have to read through everything from the beginning of the disc until you finally reach the part you want—you can simply go directly to any point and begin there.

Getting a Handle on DVD Menus

Now that we're thinking of DVDs as simply a way to store data that we can access easily, it should be fairly easy to make the jump to understanding how DVD menus work. Essentially, DVD menus are simply a means of controlling which data is accessed next when a particular button is pressed. In many ways, DVD menus are similar to the underlined links you see on Web pages. They contain the address of some other place you might want to go, and they don't require you to do anything except to click them in order to bring you there.

> **NOTE**   *In this discussion of DVD menus, most of the same things apply to VCD menus, as well. One notable exception is that VCD menu items must be selected by number rather than by being clicked. Still, DVD menus and VCD menus are more alike than they are different.*

To get a better understanding of DVD menus, take a look at Figure 8-1. Here I've added a menu to a movie that has four different scenes (the fourth one is not visible, because contracting the timeline to show it would make the chapter markers too small to be understandable in the figure).

In this case, I've included a menu button for each of the four scenes. In DVD parlance, a scene that is linked to a menu button is called a *chapter*. As the figure shows, a menu track is added to the timeline just above the main video track whenever you create a menu. That menu track includes menu markers, chapter markers, and markers that indicate a return to the menu. Each of these has some special characteristics that help make DVD menus work:

- ■ **Menu markers**   These represent a section of the timeline that uses a title overlay to display the menu over a background image. This background can be still or moving, depending on the effect you want to create. The menu includes the chapter buttons that link to the various chapters (or additional menus).

- ■ **Chapter markers**   These indicate the frame where the scene (chapter) will begin playing when the associated button is clicked. Normally, this is the first frame of the scene, but you can move the marker further into the scene if you want to trim some of the frames at the beginning of the scene. Chapter markers control the start of playback, but not the end of it. Therefore, clicking the Chapter 1 button (in the figure) will play Chapter 1 and then Chapter 2 sequentially.

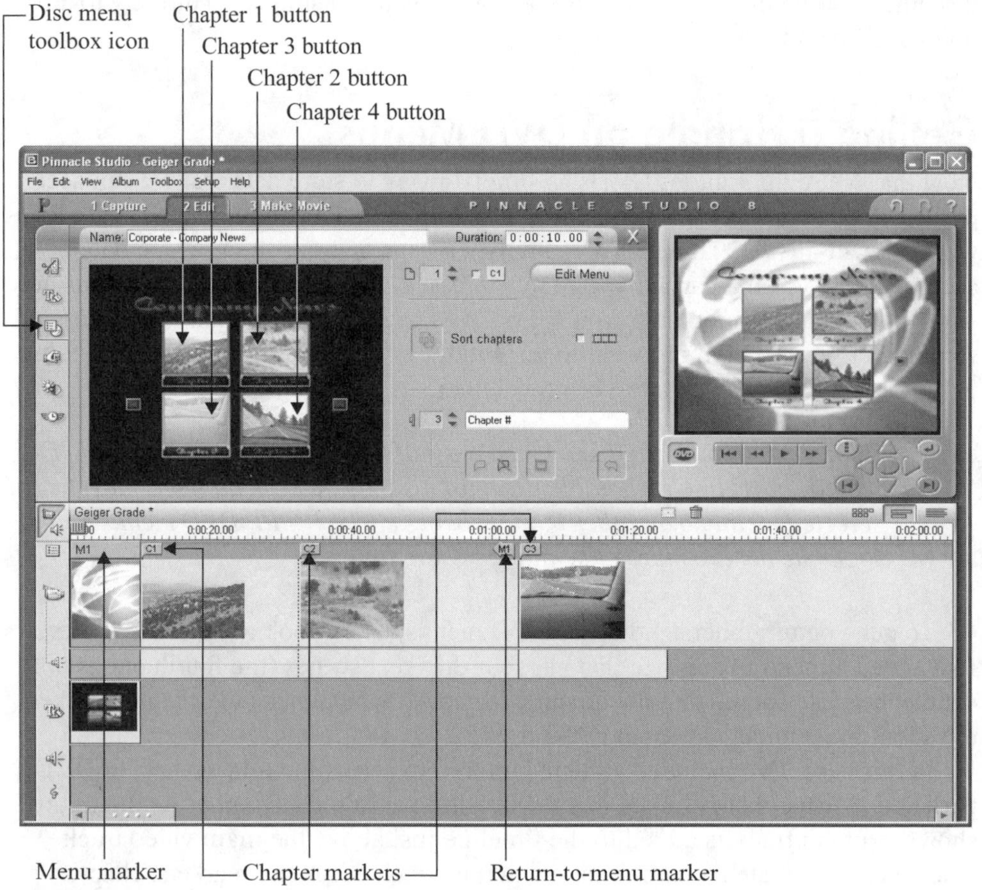

Disc menu toolbox icon

Chapter 1 button

Chapter 3 button

Chapter 2 button

Chapter 4 button

Menu marker — Chapter markers — Return-to-menu marker

FIGURE 8-1 This is an example of creating a DVD menu.

- **Return-to-menu markers** These cause a jump back to the menu, thus preventing the playback from automatically continuing through the movie. When you add one of these markers it is normally placed at the last frame of the selected clip, but you can move it if you like. In Figure 8-1, I've added a return-to-menu marker at the end of Chapter 2, so the only way to view Chapter 3 is to select it from the menu.

Believe it or not, these three markers provide everything you need to create the complex interactions I mentioned earlier. If you want to create a special section of outtakes, for example, you could place them in a chapter before or after the main part of your movie. Then, you could use return-to-menu markers to make sure that your main movie and your outtakes chapter would both return to the menu when they finished playing, ensuring that they play completely independently of each other.

DVD menus have one additional characteristic that is vital to understand. Whenever the timeline scrubber is within a section marked by a menu marker, the playback automatically loops back to the beginning of the same menu as soon as the scrubber reaches the end of that section. It's almost as though a return-to-menu marker existed at the end of the menu section—the reason one isn't used is that the accidental deletion of a return-to-menu marker would stop the menu from looping. In other words, menus always continue to repeat until you do something (like click a menu button) to move the timeline scrubber out of the menu.

Because DVD menus continually loop, you can make them as long or as short on the timeline as you like. If your menu has a moving background, the duration of the menu loop may have some effect on how cleanly the background transitions from the end of the scene back to the beginning, but frankly I doubt that most viewers would really notice this very much. On the other hand, if you choose the option to include moving videos as the thumbnails on the buttons, the duration of the menu sequence is also the duration of those moving videos on the buttons. That is, if your menu is set to ten seconds on the timeline, the buttons will continually loop through the first ten seconds of each chapter's video.

Creating DVD Menus

As I mentioned in Chapter 7, DVD titles and DVD menus are closely related. You can, in fact, use the title editor to create menus manually if you are so inclined. As a practical matter, though, I think that you will find that it really is a lot easier to use the specialized menu creation and editing tools to create your menus.

| NOTE | *Before we proceed, I'd like to point out that even though some digital video editing programs place arbitrary limits on the number of buttons you can have on each menu, aesthetic considerations will probably determine the actual number used. That is, you'll want to keep your menus fairly simple in order to make them easier to use. Packing too many buttons onto a single menu can make it difficult for someone using a set-top DVD player to navigate the menu. A reasonable alternative is to create multiple linked menus (as you'll learn how to do a bit later in this section).* |

Adding a Menu to the Timeline

Since it is far easier to use the menu creation tools than if you try to create menus from scratch, we'll use those tools to add our menu to the timeline. Figure 8-2 shows the basic steps to get you started.

Let's go through the steps to add the menu:

1. Click the Show menus tab to open the menus album. You can also use the Album/Disc Menus command to open this album.

2. Select the album style you wish to use. You'll want to watch for the symbol in the lower-right corner of the thumbnail, which indicates menus that have a moving background.

3. When you have made your choice, drag it onto the timeline and drop it. After you have dropped the menu onto the timeline, the menu track will appear.

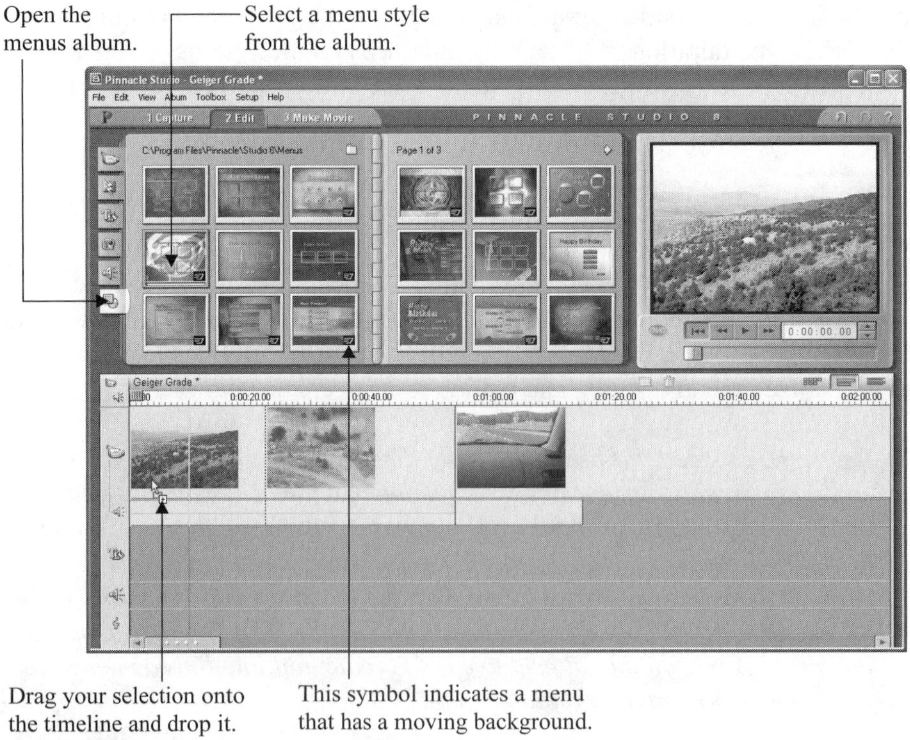

Open the menus album.

Select a menu style from the album.

Drag your selection onto the timeline and drop it.

This symbol indicates a menu that has a moving background.

FIGURE 8-2 Start with a menu from the menu album.

4. When you see the message, shown here, asking if you want to automatically create links to the scenes in the movie, click Yes. Otherwise, you will have to create the links manually after the menu has been added to the timeline.

Adding Menu to Movie

Would you like Studio to automatically create links to each scene after the menu?

☐ Don't ask me this again

Yes   No

Once you have dropped the menu onto the timeline, Studio automatically opens the video editing toolbox and displays the Create or edit a disc menu tab. We'll take a look at the options on that tab in the next section.

| TIP | *If none of the background images in the menus included in the menus album appeals to you, don't worry—you can easily change the image. It's probably more important to choose a menu whose button shapes and layout suit your needs, since those are a bit more work (although not difficult) to change.* |

Editing Your Menu

Now that you have created a basic menu, you'll probably want to make some changes to it. One of the most likely reasons you'll want to make changes is that you don't want every scene in your movie to be considered a chapter with its own button. If you allowed Studio to automatically create the links to the scenes, each scene would have its own button. You'll probably want to remove some of those extra buttons and leave only the links you really want on the menus.

In addition to removing some of those extra buttons from your menus, you may also want to edit your menus like this:

■ Add some return-to-menu markers to break your movie into several independent sections.

■ Add moving images to the buttons to replace the default of the first frame of the scene.

- Change some of the existing wording on the menus or add some additional text.

- Add some background music that will play while the menu is onscreen (we'll look at this subject in more detail a little later in this chapter).

- Change the background image or the background video that displays behind the menu.

With the exception of the menu-specific items on this list, you can probably figure out most of the methods you would use without any extra assistance. For example, adding background music is something you learned about in Chapter 6, and changing the text in titles is something you learned about in Chapter 7. The fact that you want to make these types of edits to a menu rather than to an ordinary scene or a title really has no affect on how you go about the tasks.

Figure 8-3 shows the menu toolbox, which contains the various items you'll use to do most of your menu editing.

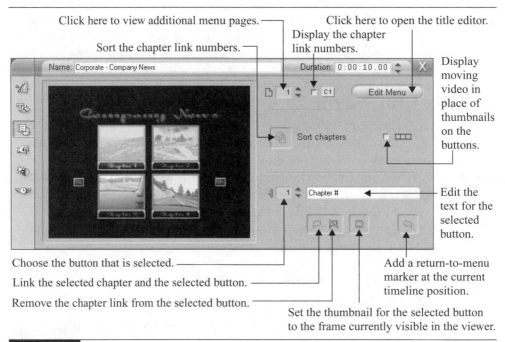

FIGURE 8-3   Use these tools to edit your menus.

Let's take a closer look at the specifics of these tools:

■ Menus often have several pages, since in most cases there are more buttons created than will fit on a single page. The page selector enables you to work on any of the pages.

■ When you have a large number of chapters and menu buttons, it can become a bit confusing trying to figure out which button is linked to which chapter. As this shows, you can choose to display the chapter numbers in the menus (these numbers will match the chapter numbers shown on the menu track):

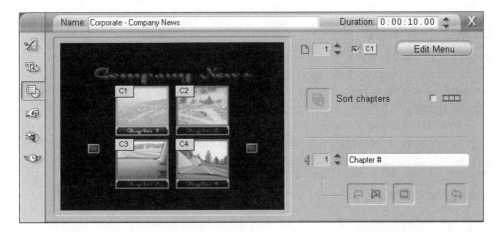

■ If you delete some of the chapter links, it's quite possible to get the buttons out of sync with the order of the chapters. You can click the Sort chapters button to again place the buttons into sequential order.

■ Menu buttons normally display a thumbnail image of the first frame in a chapter. You can use the Set thumbnail option to choose a different frame, or you can use the Display moving image option to show the first several seconds of the chapter's video on the button. Note that it may take several seconds for your system to render moving video for the button faces, so you won't be able to preview this effect until the rendering is completed.

■ By default, the text for each button shows "Chapter" and the chapter number. You can change this to something more descriptive using the text box. Keep in mind, however, that, if you create a VCD copy of your movie, viewers will be able to navigate the menus only by using the number keys on their remote. It's a good idea to at least retain the sequential numbers to make this navigation somewhat easier.

■ You can set or remove links to chapters using the two lower-left icons in the toolbox. Doing this enables you to set the specific chapter points you want linked to menu buttons and thereby have better control over your menus.

■ You can use the lower-right icon to set a return-to-menu point. Doing this prevents the DVD player from moving further along the timeline, unless you have set another chapter button to a later point on the timeline. This technique enables you to create the different special sections in a movie that can be viewed only through a menu selection (as opposed to simply having them play when the timeline reaches their position during playback).

Linking Menus

You have already seen that menus can have multiple pages. When a menu has more than one page, each page of the menu looks pretty much the same, with the identical backgrounds and similar buttons. In most cases, this is of course just what you want.

Sometimes, though, you may want to create more than one menu for your disc. Suppose you were creating a movie that included several special extras, for example. In that case, you might want to create a main menu that the viewer could use to play the main part of your movie, but you would also include a link to your "extras" menu, where they could select from all of the bonus content. By keeping the two menus separate, you would be able to make the two menus look considerably different from each other.

There is just one little problem with this scenario. As you've already seen, it's easy to link menu buttons to specific chapters, but there hasn't been any way to link to other menus. Well, never fear, the solution is quite simple.

Figure 8-4 shows an example of a movie where I've added a second menu to the timeline (remember that the menu track designates menus with a solid color band and the letter M followed by a sequential menu number). In this case, I've set the M1 menu to show a couple of teaser scenes, and I've set the M2 menu as the main movie menu. It really doesn't matter what order I use for the menus (except that the M1 menu will be displayed when the movie is played).

To link a menu button to another menu, you follow pretty much the same routine as when you are manually linking a button to a chapter:

1. Create both menus.

2. If it is not already open, you will need to open the Disc menu toolbox.

3. On the first menu, select the button you want to link to the second menu.

FIGURE 8-4   Here I've added two menus to the timeline.

4. Right-click the second menu on the timeline to display the pop-up context menu.

5. Select Set Disc Chapter, as shown here. This will add the link to the selected button.

> Delete
>
> **Set Disc Chapter**
> Set Return to Menu

When you link menus, remember to include a link back to the first menu. Otherwise, you will be creating a one-way street that will prevent your viewers from returning to the original menu. (Well, okay, they can return to it by removing the disc from their DVD player and then inserting it again—but that's kind of lame, don't you think?)

Incidentally, you'll need to add the chapter links to the second menu manually, because they are added automatically only to the first menu in your movie. As you've probably already guessed, you add those chapter links the same way you add links to other menus—except that you right-click a scene rather than a menu on the timeline.

Adding Background Music to Your Menus

There's no reason why your menus have to be silent when it's so easy to add some background music to them. By now, you probably won't be surprised to learn that adding background music to a menu is just as easy as adding background music to any other movie clip. In Figure 8-5, for example, I've decided to add some appropriate background music to one of the menus in my movie using the SmartSound feature in Pinnacle Studio.

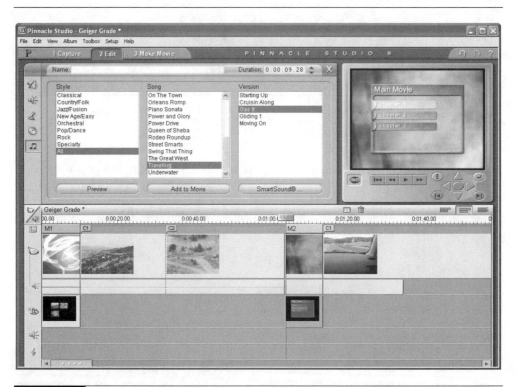

FIGURE 8-5   Here I'm adding background music to the second menu.

There is, however, an important consideration you'll want to remember when you add background music to a menu. As you'll recall, menus automatically repeat until the viewer makes a menu choice. Accordingly, the background music will also continue to repeat as long as the menu is displayed. While this isn't really a huge problem, you may find that increasing the duration of the menu on the timeline is a good idea. That way, the musical segment will have more time between repetitions of the same notes.

NOTE   *You may find that certain songs (or versions) aren't listed in the SmartSound album when you try to add background music to a menu. The reason for this is that the default length of a menu is simply too short to allow certain song versions to play. To correct this, lengthen the duration of the menu clip before you attempt to select your background music.*

VCD Menu Considerations

So far we've concentrated on DVD menus, but you can create menus for your VCDs as well. In fact, you create VCD menus exactly the same way as you do DVD menus.

There is, however, one very important difference between the way DVD menus and VCD menus function (S-VCDs work just like VCDs). When you make a selection from a DVD menu, you can move the cursor around the screen and click on a button to select it. This is not possible on a VCD menu. Rather, on a VCD menu you must make your selections by using the number keys on the remote control.

What this means is that you should include the sequential chapter numbers in the button captions if you are creating a VCD. Otherwise, your viewers will probably find that navigating your menus is difficult and confusing.

In this chapter, we've looked at just what it takes to make your movies easy to navigate by creating menus. You've learned that menus give you tremendous flexibility in laying out your movies and that you can even use menus to add some of those fancy extras that are sometimes found on commercially produced DVD movies.

In the next chapter, we'll take a look at the various ways you can output your movies. Although we've mostly been talking about producing DVDs, you have quite a few different options available to you, making it possible for you to use your videos in a number of exciting ways.

CHAPTER 9

Choosing Output Formats
for Your Movies

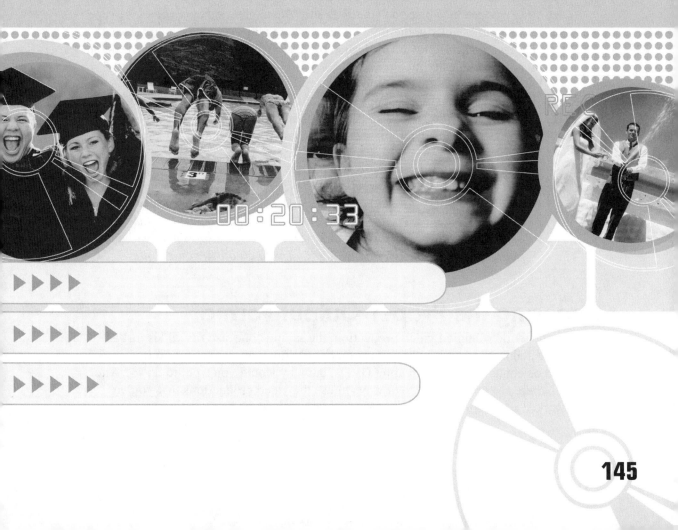

In the earlier chapters, we covered all the basics of creating your movies using a digital video editor. By now, you should be able to capture your video, select clips to include in your movie, edit the scenes, add audio tracks, create titles, and add menus to your productions. All of this has been a lead-up to actually outputting your movie so that other people can finally watch it. Now it's time to take that last step and create your finished product.

Even though we've primarily been discussing creating DVDs, in reality that is but one of the output options that is available to you. In this chapter, we'll look at those other options and we'll also take a closer look at the optional settings you can use to optimize your movies for the various output possibilities. For example, you might want to create several different versions of your video production to meet several different needs. You could create a VCD version of a member recruitment video to hand out at a fair, and you could also create a streaming video version of that same production to place on your Web site. You might want to create a videotape version of your "family gathering" movie to share with relatives who might not have access to a DVD player, while also producing your interactive DVD version for the more technically inclined relatives.

This is probably a good time to point out that your digital video editor may not offer the complete range of output options that I'll mention in this chapter. The Pinnacle Studio software I'm using to demonstrate the movie-making fundamentals just happens to offer quite a complete set of output options.

> **NOTE** *Once you choose your output format and options, your digital video editing program will still need to* render *the file to prepare it for use. You start this process by clicking the Create button. Depending on the speed of your PC, the format you have selected, and the length of the movie, this task may take a considerable amount of time. You'll have to experiment with your system to determine just how long it takes to create the output file.*

Choosing the Proper Output Format

In the world of digital video production, the saying "one size fits all" is just as lame and bogus as it is in regards to clothing. Sure, you can accept limited choices and be content with just creating DVDs, but why would you want to do so? After all, there are so many different ways to use the video productions that you've worked so hard to perfect.

Different uses call for considerably different output settings. DVDs can offer very high quality audio and video because a DVD has plenty of room to store a lot of data. On the other hand, a streaming video production that you want to include on a Web site must sacrifice both audio and video quality in order to keep the amount of data low enough for dial-up access to be a reasonable option—it's true that you can ignore dial-up users and just aim your site at broadband users, but in doing so you'll be cutting your potential audience by at least two-thirds.

In the following sections, I'll discuss both the various options that are available for each type of output and the reasons why you might want to choose a particular output format. You'll probably find that some of the output formats really do offer some very compelling reasons why you may want to choose them. In showing you the options, we'll use the Pinnacle Studio Setup Options dialog box, which appears when you click the Make Movie tab in Studio and then click the Settings button.

TIP	*Always save your project before you create your final movie file. Rendering the movie file should go without a hitch, but this task is probably the most computing-intensive part of the whole video production process, and therefore it is also where problems are most likely to occur. By saving your project first, you insure that you will be able to recover all of the work you've put into your movie in the event something does go wrong at this stage.*

Writing to Tape

The first output medium we'll have a look at is tape, which can include both analog and digital tapes. A standard VCR, for example, is an analog tape, while writing the movie back to your DV camcorder is an example of digital tape output. As far as your digital video editor is concerned, there is very little difference between the two. In fact, the only functional difference is that your PC can operate the record and stop functions on a DV camcorder, while you'll have to operate these controls manually if you're outputting to a VCR.

NOTE	*The IEEE-1394 (FireWire) connection between your DV camcorder and your PC is a two-way connection. This is not the case for the analog connections to a VCR. If you want to record your movie on a VCR, you'll need to connect the composite video or S-Video output on your PC to the video input on your VCR. Likewise, you'll need to connect the audio outputs from your PC to the audio inputs on your VCR. Remember, composite video and S-Video connections are generally not found on PCs, so you'll need an analog capture device with both inputs and outputs if you want to write to a videotape in your VCR.*

Let's consider a few reasons why you might want to output your movie to tape:

■ A videotape can be played in any standard VCR. This makes this output option desirable for sharing your video with people who do not have a set-top DVD player. Keep in mind, however, that videotapes will not include any menus you may have added to your movie, so you may need to modify the layout of the movie so that it can be played back in a totally linear fashion. That is, the movie must be set up so that it can be played from beginning to end without relying upon any fancy tricks. If, for example, you have an outtakes section, you'll want to place it after the end of your movie so that your movie can be played completely through without interruption.

■ Virtually everyone knows how to insert a tape into a VCR and play it. Therefore, a videotape can be good for sending a copy of your movie to someone who is a technophobe or who simply can't read and follow simple directions (or for someone who can't understand why they can't play a DVD movie in their audio CD player).

■ Most DV camcorders can play back movies on a standard TV set. If you are making a movie that you will use as part of a presentation, your DV camcorder is almost certainly a lot easier to carry along than a VCR. You may not be able to count on finding a set-top DVD player at the location where you will be speaking, so carrying along your DV camcorder with your recorded movie can be a good way to insure that you will be able to show your video presentation. Be sure to remember to bring along the plug-in power supply, though, so you don't have to worry about running out of battery power in the middle of your talk.

■ Since mini DV tapes reproduce the movie at full quality, recording your movie on your DV camcorder can be an excellent method of sharing the video production tasks with someone else. For example, if you shot a live music video for a local band, they might want to have a studio recording of their songs added to the movie to replace some of the live recordings. Or perhaps you're working with a wizard at sound effects on the other side of the country who you would like to have add some special sound effects to the video portions of your movie after you've finished editing. Sending the mini DV tape for additions would be a whole lot cheaper than traveling to get you both at the same location.

Now that you've seen some of the uses for tape, let's take a look at the options that are available for writing to tape.

Writing to DV Camcorders

Here you can see the Make tape tab of the Pinnacle Studio Setup Options dialog box:

In this case, I've selected DV Camcorder as the playback device from the drop-down Video list box. When you make that selection, the DV Camcorder is automatically selected as the audio playback device (you don't have the option of making a different selection, because the audio playback device is always set according to the video playback device).

Because DV Camcorder is the video playback device selection, you also have the option of selecting the "Automatically start and stop recording" checkbox. It's a good idea to select this option, since then your PC can control the camcorder so that it starts and stops recording at precisely the correct points in your movie. If you don't select this option, you'll need to manually control your camcorder in much the same fashion as if it were an analog device such as a VCR.

You can also choose to delay the recording slightly using either the seconds or the frames settings. You might want to do this if your camcorder has problems stabilizing the tape when a recording begins. Increasing the delay time actually starts the camcorder earlier, because the delay refers to how long after the camcorder begins rolling that the video signal is sent to the camcorder. You may need to experiment a little here to see what setting is needed for your camcorder. If the tape that your

DV camcorder records seems to be missing some of the early frames in your movies, increase the delay time to compensate. On the other hand, if the tape starts out with a still image of the first frame of your movie, the camcorder began recording too soon, and you'll need to reduce the delay time slightly.

Writing to a VCR

Writing to a tape on a VCR is a more complicated project than writing to a tape in your DV camcorder. The deceptively simple appearance of the options shown here really doesn't give much of a clue about what you're up against. In fact, you may need to practice the whole process a couple of times before you get the results you really want.

To begin with, notice that the video playback selection is shown as the VGA Display. The reason for this is that you need to have analog capture hardware installed that will send the video signal from your screen out through either composite video or S-Video jacks. Next, your VCR must be connected to one of these jacks as well as to the audio output jacks on the capture hardware. To actually make your recording, you must have the VCR ready to record, hit the Create button on the Make Movie tab, and then click the Play button on the viewer at the same time that you hit the Record button on your VCR. It helps to have everything set up and ready to go ahead of time. You'll also soon discover that knowing precisely which buttons need to be pressed (without a lot of fumbling around trying to figure out which button starts the recording) will be a big help!

> **TIP** *If you're recording your movie on a VCR, it may be a good idea to pad the beginning of your movie with several seconds of black screen. You can do this by adding a blank title to the beginning of the main video track. That way, it won't be quite so critical that you hit the Play button on your PC and the Record button on your VCR at precisely the same time. Alternatively, you could add an ordinary title at the start of the main video track and then set it to display several seconds longer than you might otherwise do. The whole idea here is to make certain that none of the motion in your video is missed when playback begins.*

Creating AVI Files

AVI *(Audio Video Interleaved)* files are intended for playback on a PC rather than on any type of set-top player (in fact, the only place an AVI file can be played is on a computer). As is shown here, AVI files have quite a few optional settings that you can use to adjust the playback quality and file size:

Pinnacle Studio Setup Options

Make tape | Make AVI File | Make MPEG File | Make RealVideo® | Make Windows Media® | Make disc

Video settings

☑ Include video in AVI file
☐ List all codecs

Compression: [Options...]
Indeo® video 5.10

Width: 352   Height: 240   Frames/second: 29.97

⦿ Quality   ◯ Data rate   Percent: 100

Audio settings

☑ Include audio in AVI file
☐ List all codecs

Type: PCM

Channels: 16-bit stereo   Sample rate: 16 kHz

Force settings to be same as current project
[Same as Project]

[OK]   [Cancel]   [Help]

AVI files have many uses. These range from training videos intended for playback on a PC to a video file on a Web site. Unlike the streaming video I'll mention later in this chapter, however, AVI files must be downloaded completely before playback, which limits their popularity for Web-based video considerably. This is not to imply that you can't use AVI files as a part of your Web site, it's just that you should be aware that visitors will have to wait for a long download to complete before they can begin viewing your movie.

> **NOTE** *AVI files use* codecs *(compression/decompression methods) to reduce their size by compression when the file is created and decompression when the file is played back. Since the same codec must be used for both compression during recording and decompression during playback, selecting one of the more obscure codecs can limit your potential audience, since it's less likely that your intended viewers will have the correct codec installed on their PC. For that reason, it's usually best to stick with the default codec selections, since they tend to be the more popular (and therefore widely distributed) codecs—you're less likely to cause playback problems from not having the proper codec available.*

Choosing Your Video Output Options

The video output options have the potential for affecting to a considerable degree both the quality and the size of the resulting AVI file. Therefore, it is quite important for you to understand exactly how each of the options influences these factors. AVI files are generally rather large—even if you have been quite careful in selecting options—so you'll want to make careful choices.

- The "Include video in AVI file" must, of course, be selected if you want your AVI file to include the video portion of the signal. Deselecting this option produces a file that contains only the audio portion—this is something you might want to do in very rare instances, such as when creating a compressed audio track.

- It's best to leave the "List all codecs" option deselected to prevent you from inadvertently choosing a codec that viewers might not have on their PC. If, on the other hand, you are producing a video for someone who has specifically requested that you use some obscure codec, selecting this option makes it more likely that you'll find that codec in the Compression list (but you'll find it only if it is installed on your PC). If you need a specific codec that is not installed on your PC, you'll probably have to

contact the manufacturer of that codec to determine how to obtain a copy
that you can install on your system.

■ Clicking the Options button displays a dialog box with the optional settings
for the currently selected codec. The dialog box will have the settings specific
to just that codec, so the dialog box will look somewhat different depending
on which codec is currently selected.

■ The Compression drop-down list box enables you to select a specific codec
to use to compress your movie. Different codecs vary in both their ability
to reduce the size of the AVI file and in the final quality of the file. You can
experiment to see which codec works best for your needs, but it is generally
safest to use one of the Indeo video codecs (which were created by Intel),
since they seem to be most widely available and therefore most likely to
work on the broadest range of systems.

■ The Width and Height boxes enable you to specify the resolution in pixels
for the video portion of the AVI file. Reducing the width and height has a
very large effect on the file size (as well as on the ability of slower PCs to
display the video without dropping frames). The reason for this is simple—
the total number of pixels in any one frame is equal to the width in pixels
times the height in pixels. If you cut both the width and height in half, you
reduce the total number of pixels in each frame to one-fourth the original
value (for example, .5 times .5 equals .25). Multiply this difference by the
total number of frames in your movie and it's easy to see why reducing
the width and height has such a huge effect on the overall file size.

■ The Frames/second list box enables you to choose a slower frame rate to
further reduce the file size and the demands on the processor of the system
displaying the video. The standard rate of 29.97 frames per second (fps)
produces the best flicker-free video, but reducing the frame rate is often
a good option for videos that will be downloaded or played on slower
computers. Keep in mind, however, that lower frame rates will produce
a jerky effect if your video includes any fast motion, since there will be
larger differences between each successive frame. Anything below about
15 fps tends to look pretty jerky in the final video.

■ The Quality/Data rate slider specifies the amount of compression that is
applied to the file (you will have either the Quality or the Data rate option,
depending on which codec you have selected). At 100 percent, the video
quality is the best but the file is largest, since virtually no compression is
applied. At lower settings, the quality suffers but the file size is smaller,

since higher levels of compression are applied. You may need to experiment to find a setting that produces satisfactory results. To do so, create a couple of different AVI files using different settings of this slider, then see which one produces acceptable video quality along with the smallest file size. Keep in mind, however, that the PC playing back the movie has to decompress the file during playback, and slower computers may not be able to handle extremely high compression ratios very gracefully.

Choosing Your Audio Output Options

The audio output options are generally somewhat less important than the video output settings, but they both do have an influence on the size and quality of your AVI file. Let's take a look at each of the audio output options:

- Unless you intend to produce a silent video, you should leave the "Include audio in AVI file" checkbox selected. Deselecting this option does reduce the size of the AVI file, of course, so it may be appropriate to do so for videos where the audio track adds little or nothing of value.

- As with the same setting in the video options, it's best not to select the "List all codecs" option in the audio settings. Select this only if you have the specific need to use an audio codec that is not shown in the drop-down Type box (and you know that you have the desired codec installed on your PC). Remember that the computer playing back your AVI file will also need to have this same codec installed.

- Use the drop-down Type list box to select the codec for compressing the audio. Generally, it's best to use PCM (Pulse Code Modulation) or ADPCM (Adaptive Delta PCM), since these are the most common codecs and are installed on most PCs. Some other codecs produce higher compression ratios and therefore smaller files, but using them comes with the increased risk that other people won't have the correct codec installed.

- Use the Channels drop-down list box to select stereo or monaural sound and to select the data width. Unless your video includes a lot of music, you can probably get by with 8-bit mono, which will reduce the file size considerably. Remember that most camcorders use a single microphone, so selecting stereo sound simply increases the file size without adding any benefit (if you are using the recorded soundtrack). In reality, even videos that include background music often don't need to have stereo sound— especially if the music is a minor part of the whole production.

■ The sample rate setting determines the highest-frequency sounds that will be accurately reproduced. Generally speaking, the sample rate should be twice the highest frequency that you want included. Lower sample rates produce smaller file sizes, and even an 8 kHz sample rate is perfectly adequate for reproducing most speech. You may need to try a couple of different settings to see which ones works best for you.

NOTE	*Although the Same as Project button is on the same side of the dialog box as the audio settings, clicking this button affects both the audio and video settings. If you have been creating a DVD movie, clicking this button will result in an AVI file with the highest quality settings possible—and the largest AVI file, too. Unless you really need full DVD quality in your AVI file, it's best to avoid clicking this button.*

Creating MPEG Files

MPEG (Motion Pictures Experts Group) files are files that can be used for quite a few different purposes. These include anything from putting them on a Web site to creating full DVD quality files for later burning to disc. MPEG files are always compressed, but you can fine-tune the settings to get just the right balance of quality and file size.

This shows the Make MPEG File tab of the Pinnacle Studio Setup Options dialog box:

In this case, I've selected the Custom option in the Presets drop-down list box so that I can have the full range of adjustments available. Selecting any of the other presets (such as Internet Low Bandwidth or DVD Compatible) grays out all of the options on this tab and simply applies a set of predefined settings.

Selecting the Video Settings

The video settings section contains a number of important options that can have a very broad range of effects on both the video quality and the size of the file. Let's take a close look at each of these settings:

- The choice of MPEG1 or MPEG2 for the compression method is extremely important. MPEG2 creates higher quality video, but it also requires that the PC that plays your movie have an MPEG2 decoder program installed. This is typically not a problem with the newer PCs, but it might be a problem if your target audience is using older systems. Note that MPEG2 is the compression method that is used on DVD movies, so that should be your choice if you intend to burn the file on a DVD eventually. MPEG1 files are limited to lower resolution settings than MPEG2 files are, so this is also an important consideration in choosing between the two options.

- It should be obvious: unless the Include video option is selected, your MPEG file will not contain the video portion of your movie.

- You can use the Filter video option to smooth the edges of moving objects in your movie. Unfortunately, selecting this option tends to reduce the overall video quality (by making it appear as though the camera was slightly out of focus). This is one of those cases where you'd like to have both the smoother edges and the higher video quality, but you can only choose one or the other.

- The Draft mode option should not be used for the final output, because the quality is considerably lower in this mode. You're likely to use this mode primarily to get an overall feel for how the movie will look and feel (ignoring the quality aspect, that is). The main reason for using this mode is that rendering will be much faster, so you'll be able to test the menus and timings without waiting for a high-quality rendering to finish. Be sure to deselect this option before you create your final file.

- The Width and height drop-down list box lists all of the resolutions that are available at the current compression setting. Higher settings produce both higher quality video and, of course, larger file sizes. The same types of file size differences apply here as they do with the similar settings I discussed earlier in the section on AVI files (although the effects of MPEG compression do tend to mask the size differences somewhat).

■ The Data rate slider essentially sets the level of data compression that is applied to the movie. Higher data rates produce better quality video and large file sizes. This option can be useful if you want to squeeze just a few more minutes onto a disc, because you can reduce the data rate slightly to create a smaller size file than if you use one of the presets (which do not allow you to tinker with the data rate).

Selecting the Audio Settings

The audio settings for MPEG files are actually fairly simple. They do affect the file size, of course, but far less than do the video settings. Let's take a look at the audio settings:

■ The Include audio setting is fairly self-evident. If you don't select this option, your movie will be silent.

■ The Sample rate setting is similar to the same setting for AVI files, except that only two options are available: 44.1 kHz (the standard for audio CDs) and 48 kHz (the standard for digital video). Since the two options are so similar, you probably won't notice much (if any) difference between the two. Incidentally, you don't have to use 48 kHz simply because it is the standard setting for digital video—your movies will play just fine either way.

■ The Data rate slider is similar to the video data rate slider. Higher data rates produce higher audio quality at the expense of larger file sizes. Here, too, the choice of an optimal setting is probably best determined through experimentation. Try out several settings until you find one that produces acceptable sound quality along with reasonable file size.

Creating Streaming Video

Streaming video is the term used to describe movies that are primarily intended for playback by people who visit a Web site. Unlike an AVI file or an MPEG file, which must be completely downloaded before playback can begin, streaming video begins playback shortly after the download starts. As the movie plays, the data continues to download in the background. If there are no unanticipated delays in the download, the movie plays back smoothly and much sooner than if the entire file needed to be downloaded before the playback could start. In the real world, of course, the Internet is often a lot slower than we would like, so streaming video tends to be less satisfying than it should ideally be.

NOTE	*There are two primary types of streaming video that are most popular today:* RealNetworks RealVideo *and* Microsoft's Windows Media *both have their fans and detractors. RealNetworks RealPlayer G2 player has certainly been around longer than the Windows Media Player, but unfortunately RealNetworks alienated quite a few users when it became known that they were collecting data about users without prior approval or notice. As a result, Windows Media Player has become at least as popular as the RealNetworks RealPlayer G2 player among many Internet users. You will have to decide for yourself which format is best for your needs (although choosing* Windows Media *does leave out the very small Macintosh and Linux audiences, since they do not have the Windows Media Player).*

Creating RealVideo Files

If you choose to create RealVideo files, you will have a few relatively simple options, as shown here:

Here is a brief explanation of these settings:

■ The Title, Author, Copyright, and Keywords text boxes enable you to
enter the appropriate information (specific to your movie file) in each
of these boxes. You can use any or all of them, but their use is optional.
I recommend filling in both the Author and Copyright boxes, however,
so that your movies will at least contain information to prove that they
are yours. Place the current year in the Copyright box. The Title, Author,
and Copyright information is encrypted so it cannot be easily viewed or
changed by the casual user.

■ The Video Quality drop-down list box enables you to choose the quality
level you prefer. You may need to do some experimentation to determine
which setting works best with the type of video that is contained in your
movie. Selecting a sharper image tends to make rapid motion somewhat
jerky, while selecting smoother motion tends to reduce the quality of still
images.

■ The Audio Quality drop-down list box contains several choices, which
are pretty self-descriptive. For example, the Voice with Background Music
selection produces a slightly larger file with higher audio quality than does
the Voice Only selection.

■ The Video size selections specify the size of the video window. Smaller
sizes produce smoother downloads with low-speed connections, but larger
windows are always easier to see. Remember that the same issues of video
size we've discussed earlier apply here as well.

■ Unless you have access to a RealNetworks RealServer, you'll want to choose
HTTP for the Web server. When you choose HTTP, you can simply place
the RealVideo file on your Web site with a link that visitors can click to
view your video. The RealNetworks RealServer does have the advantage
of enabling you to produce RealVideo files for several different target
audiences at the same time (since the RealNetworks RealServer detects
the connection speed and supplies the correct version to suit the visitor).

■ The Target audience selections specify the connection speed at which the
file is intended to be viewed. If you select the HTTP Web server option,
you will be able to make only a single selection in this area. If you select
the RealNetworks RealServer option, you can choose as many selections
as you like but the size of the file you'll need to upload to the server will
increase, because there will be multiple versions of your movie within the file.

TIP

Even if you select the HTTP Web server option, it is still possible to accommodate visitors with different connection speeds. Simply create two (or more) files targeted at a specific speed, then offer your visitors a couple of choices using different links on your Web site.

Creating Windows Media Files

The settings that are available if you have decided to create Windows Media files are actually quite similar to those for RealVideo files. This shows the Make Windows Media tab of the Pinnacle Studio Setup Options dialog box:

In this case, I've filled in some of the optional items to give you a better idea how to use them.

We'll now examine the Windows Media file options:

■ The Title, Author, Copyright, and Description text boxes should be easy enough to figure out from their titles. The Rating text box enables you to specify a suggested audience rating. As in RealVideo files, the Title, Author, and Copyright information is encrypted within the Windows Media file to protect the information from being easily viewed or changed.

- The Markers for Media Player "Go To Bar" section enables you to have markers automatically placed in the file so that someone viewing your movie can more easily navigate through your movie. Essentially, these markers function quite similarly to the chapter markers in a DVD movie, and they allow someone viewing your movie to jump to specific points in the movie.

- The Playback Quality options enable you to choose the quality (and therefore the file size) of your finished movie. The left-side drop-down list box enables you to choose from three different presets, or you can choose the Custom option. If you choose Custom, you then have the option of choosing a specific target audience using the right-side drop-down list box (this list box is available only when you choose Custom). Although you cannot choose specific width, height, and audio options, choosing one of the download speed options does apply settings designed specifically for that speed choice.

Burning Discs

We're finally down to the option that you're most likely to use—creating discs from your movies. Discs are the only output option that take full advantage of all of the features of your digital video editing program, and they also offer (in the DVD format) the highest quality playback possible.

In order to create your own discs, your computer must be equipped with a DVD-R/RW, DVD+R/RW, or a CD-R/RW drive (this last will limit the types of discs you can create). A standard CD-ROM or DVD-ROM drive is a read-only device and so cannot be used to write to the recordable discs. If you are still unclear about this topic, I suggest that you go back and look over Chapter 1 again. In that chapter I discussed the various disc format issues that you must understand before you purchase blank discs to record your movies.

TIP	*If you are in the market for a DVD-R/RW drive, you may want to read Appendix A before you make a choice. There, I discuss a brand-new drive from Panasonic that can serve all of your disc writing needs as well as provide an excellent solution to the problem of hard disk space vanishing as a result of the demands of digital video storage and editing.*

Here, I've opened the Make disc tab of the Pinnacle Studio Setup Options dialog box:

In the following sections we'll discuss the various options this tab contains.

Choosing an Output Format

You can create three different types of discs, and each has certain advantages as well as disadvantages. Let's take a closer look at each of these three disc formats:

- VideoCDs (VCDs) have the lowest quality playback of any of the three disc formats. VCDs are discs that you create using CD-R (or CD-RW) recordable discs. A VCD uses MPEG1 compression, which allows for about 63 minutes of video on a 650MB disc (or 68 minutes on a 700MB disc). A VCD can be played on most PCs and on many set-top DVD players. It's important to remember, though, that a lot of set-top DVD players will not play VCDs.

- S-VCDs are a step up in quality from VCDs. Because S-VCDs use MPEG2 compression (the same compression method used on DVDs), they offer near-DVD quality from a CD-R/RW disc. An S-VCD can hold from 32 to 39 minutes of video on a 650MB disc, or 34 to 42 minutes of video on a 700MB disc. Unfortunately, the somewhat hybrid nature of S-VCDs means that they tend to be somewhat less compatible than the other two disc formats, especially in set-top DVD players.

- DVDs are the king of the discs in both quality and in storage capacity. Depending on the settings that you choose, a DVD can hold between 54 minutes and 160 minutes of video. In addition, the menus you add to DVDs are far easier to use than menus on the other two disc formats, since the viewer can navigate by pointing and clicking on a DVD (the other two require that you make selections using the number keys on the remote control). Recordable DVDs do cost more than CD-R discs, of course, but the cost per disc is coming down rapidly—at least in the DVD-R format, that is.

NOTE	*If you plan on using a game console such as the Microsoft Xbox or the Sony PlayStation2 to play back your discs, don't even think about VCDs or S-VCDs. Neither of these consoles recognize these formats, which means that they will not play the discs.*

When you make your disc format selection, the remaining options on the Make disc tab are enabled or disabled to reflect the settings that are appropriate for the selected format. Therefore, depending on the disc format you've selected, some of the options I'll be discussing in the next sections may or may not be available.

Selecting the Quality Settings

As I noted earlier in the discussion of disc formats, the disc format you select has a great influence on both the quality of the video and the amount of video which you can store on a disc. If you select either S-VCD or DVD as the output format, you have some additional quality setting options which can further modify the amount of video that a disc can hold. Let's examine those options:

- You can choose Automatic to use the default quality/disc usage settings. This setting is essentially the equivalent of the Best video quality setting, except that Automatic allows you to choose the Draft mode option (which I'll discuss shortly).

- If you choose the Most video on disc option, you can store more video on the disc, but at the expense of quality. On a DVD, this option more than doubles the potential length of the movie that you can store on a disc. On an S-VCD, the difference in length is not nearly as impressive.

■ You may also want to give the Custom option a try. This option enables you to select a data rate from the drop-down Kbits/sec list box, which makes it possible to choose a video quality/video length compromise somewhere between the extremes of best video/most video on disc. If you need a little more length than the best video setting provides, but you don't want to reduce the quality all the way to the Most video on disc setting, the Custom option gives you the means to make that selection.

■ The Filter video option is the same option I mentioned earlier in the discussion on MPEG files. When this option is selected, the edges of moving objects appear smoother, but the overall video sharpness is reduced slightly. This option has no affect on the length of the movie you can store on a disc.

■ The Draft mode option was also mentioned earlier. You use it when you want to create a test copy of your movie so that you can verify that your menus are functioning as expected. In Draft mode your movie is rendered at a lower quality level, which means that the rendering takes less time.

■ The MPEG audio option is available only if you are creating a DVD (it is used automatically on S-VCDs). Selecting this option greatly increases the length of the movie you can store on a DVD, since the audio portion of the movie as well as the video portion is compressed. Normally, the audio is stored uncompressed. Depending on your movie, the quality of the playback equipment, and the environment in which the movie is played, you may not notice any difference in the sound quality. On the other hand, some audiophiles claim that any type of audio compression creates *artifacts* in the sound that they can hear. I frankly doubt that most people who watch your movies will be able to tell the difference, and choosing this option frees up so much disc space that you can often choose a higher video quality setting. For example, at the Best video quality setting, choosing MPEG audio increases the disc capacity from 54 to 63 minutes. At the Most video on disc setting, the disc capacity goes from 114 minutes all the way up to 160 minutes!

Choosing Your Burn Options

Once you have selected the output format and the quality settings, you are ready to choose your burn options. You can choose these options only if you are creating a DVD, however, since both VCDs and S-VCDs are always burned directly to the disc.

As you can imagine, the Burn directly to disc option makes a copy of your movie on a recordable disc. With a DVD, you may want to create the movie file and test it in a DVD player application on your PC before committing to using a disc. If so, you can choose the "Create disc content but don't burn" option. This creates a file on your hard disc that is identical to the one that would be added to the recordable DVD.

The third burn option, "Burn from previously created disc content," becomes available once you have used the second burn option to create the file without burning a disc.

> **TIP** *Use the "Create disc content but don't burn" option when you want to make multiple copies of the same disc. You will be able to burn the copies faster this way, since your PC will have to render the movie only when it first creates the file on your hard disk. When you later burn the copies, the rendering will already be complete, and your discs can be burned at maximum speed without further rendering delays.*

Selecting the Media and Device Options

The selections in the Media and device options section are the final options you'll need to make before creating your disc. In reality, you probably won't have too many choices available in this section. Let's look at the options:

- In the Media drop-down list box, you'll have only a choice between 650MB and 700MB discs if you are creating a VCD or S-VCD. If you are creating a DVD, the only choice is recordable DVD media that is compatible with your DVD burner.

- In most cases, you will have only a single choice in the Disc writer device drop-down list box. The only exception would be if you have more than one writable drive installed in your PC.

- In the Copies drop-down list box, you can choose the quantity of discs you want to create (from 1 to 99). Depending on your system's performance, you may find that it is faster to burn multiple DVD copies by first creating the content on your hard drive and then burning the copies (as I mentioned earlier).

- The Write speed drop-down list box enables you to select the speed at which the disc is burned. Generally, you'll want to choose the highest value here to reduce the length of time it takes to burn a disc, but you may need to reduce this if you discover that your system cannot burn discs successfully at the higher speed. You're more likely to have problems burning discs at higher speeds if your system is slow or if you try to use it for other purposes while you are burning discs.

Burning That Disc

Once you have made all of your selections that will affect your final output, you can click the OK button to close the dialog box and return to the Make Movie tab.

Next you must make certain that you have selected the correct output format from the buttons along the left side. As shown in Figure 9-1, I've clicked the Disc button and selected DVD in the Pinnacle Studio Setup Options dialog box. At this point, you will be able to see how much space your selections will use.

If you are satisfied with all of your selections, click the Create disc button to begin the final output process (make certain that you have inserted the correct type of disc in the drive first, of course). At this point, you might as well go and get a cup of coffee or a soda, because your PC must first render the movie in the proper format and then burn the disc (or create the other type of output you have specified). This can take some time, but it's impossible to make an estimate of just how long without knowing the length of your movie, the speed of your PC, the output format

FIGURE 9-1 We are finally ready to burn the disc.

you've selected, and so on. It's best to just relax and wait, because pretty soon you'll have that fancy new disc with your own movie.

In this chapter, you've learned about the various movie formats you can create from your videos and about some ways that you can use the resulting output. You've seen how the different settings can affect the quality as well as the size of your movie files. And you've finally reached that final step of clicking the button to actually burn your movie.

Now we're going to turn our attention to creating an entire movie in each of several different digital video editors. You'll be able to take the basics that you've learned so far and see just how to apply them to your favorite video editor. To make things a bit easier for you, each of the remaining chapters will focus on just one digital video editor. That way, you can see exactly what steps to take on your system.

I'm going to suggest, however, that you at least look through the chapters on video editors other than the one you already have. By doing so, you may find that a different editor would really suit your needs better than your current one.

PART III

Creating Your Movies

CHAPTER 10

Creating a Movie in Sonic MyDVD

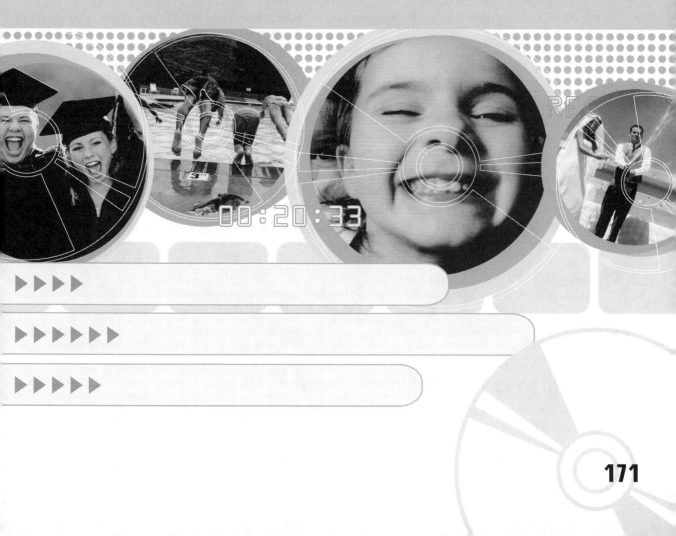

Now that you have a pretty good understanding of the basics of creating DVD movies, we're going to turn our focus in a slightly different direction. Starting in this chapter, I'm going to show you how to create a movie from beginning to end using one of the popular DVD creation programs. If it happens that a chapter covers the application that you currently have installed on your PC, you'll be able to use the chapter as a tutorial and learn what you need to know in order to use that program more effectively. On the other hand, you can use the chapters that cover some of the other programs to gain a better understanding of just how those other applications differ from the one you currently use. In that way, you'll learn which programs you might want to consider as an upgrade if you feel that you want to add additional functionality.

We're going to start out this series by covering Sonic MyDVD. The OEM (Original Equipment Manufacturer) version of Sonic MyDVD is the DVD creation software that is most often bundled with equipment such as camcorders and DVD burners, so it seems like a good idea to start our coverage with this particular program. Note, however, that I'll be covering Sonic MyDVD Video Suite Version 4.0— your bundled copy may be an earlier version or the Plus version, which may not include the ArcSoft ShowBiz video editor that is included in the Video Suite version of Sonic MyDVD. (I discussed ways you could upgrade your version of Sonic MyDVD in Chapter 2.)

By the way, I'd like to point out one thing right now. Creating your videos should be fun, rather than something that you consider to be a chore. No one expects you to turn out Hollywood blockbusters with your camcorder and your PC. Sure, you'll want to create the best looking movies possible, but be sure to enjoy your videos for what they are. Your early efforts may not be perfect, but that's just a part of the learning process. You'll get better with a little practice, and as long as you're having fun, the rest will come along in time!

TIP	*If you see features in this chapter that interest you (but might not be included in your copy of Sonic MyDVD), be sure to have a look at the back of the book for the special upgrade offer on Sonic MyDVD Video Suite that we have arranged for our readers.*

Starting a Project

Sonic uses the term *project* to describe the process of creating a movie in Sonic MyDVD. When you first open the program you will see the window shown in Figure 10-1, where you can choose the type of output you want and the type of project you are creating.

First select the output type here.   Next select the project type here.

Click here to open Sonic MyDVD directly in the future.

FIGURE 10-1   Begin by selecting the output and project types.

To start off your project, follow these steps:

1. First, select the type of output you want from the selections along the left side of the Welcome window. Your choices are DVD-Video, Video CD, and Edit Video (this third choice is available only if you have the ArcSoft ShowBiz video editor installed—either as a part of Sonic MyDVD Video Suite Version 4.0 or separately as an upgrade to the basic version of Sonic MyDVD). As you highlight each of these three choices with the mouse pointer, notice that the descriptions (and the choices) at the right side of the Welcome window change to reflect the currently selected choice. For our purposes here, make sure that DVD-Video is highlighted before you move on.

2. Next, choose the project type from the options in the right side of the window. Click Create or Modify a DVD-Video Project to begin a new movie (we'll assume that you want more control over your movie than the second option provides and that you don't already have a project you could open using the third option). When you click the option, the Welcome window will close and the Sonic MyDVD window shown here will open, so that you can begin creating your new movie:

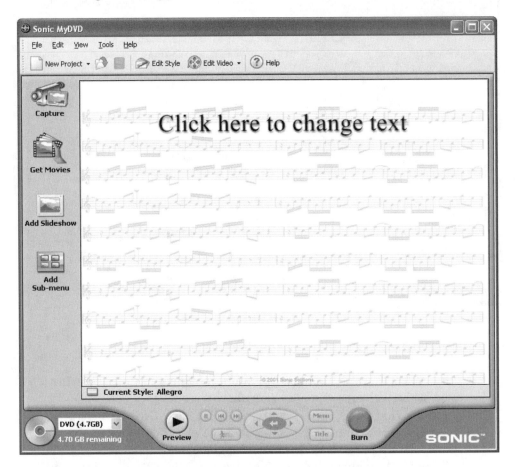

<table>
<tr><td>TIP</td><td>*If you choose to not have the Welcome screen displayed when you start Sonic MyDVD, you can always use the File/New command in the main program window to choose the type of project you want to create.*</td></tr>
</table>

Importing Your Video

The next thing you need to do is to import your video. Here you have three options to suit different needs:

- You can capture new video content directly from your camcorder.

- You can use existing video that you've captured earlier.

- You can use still images to create a slideshow.

For our movie project, we'll capture some new video content, but we'll also have a look at those other two options, since you're probably going to find uses for them in the future even if you don't use them right now.

> TIP
>
> *Remember, you can include more than one type of content on the same disc. You could, for example, include the images from your digital camera in a slide show in addition to the video from your camcorder.*

Capturing Video

When you want to use new video content from your camcorder, you must first capture that video on your PC's hard drive. Depending on the type of camcorder you have, this could be something that you'll do using an IEEE-1394 cable (for a digital camcorder) or using an analog video capture board (for an analog camcorder). If you need more information about this equipment, you may want to review Chapter 3 before you continue.

Figure 10-2 shows the Sonic MyDVD Capture dialog box that you use to capture your video and audio from your camcorder. You'll use this same dialog box regardless of whether your camcorder is digital or analog.

Let's take a look at exactly how you go about capturing your video in Sonic MyDVD:

1. Connect your camcorder to your PC and get it ready to download the video. You will have to turn on your camcorder, and you will likely have to set it up for playback. On my Sony DV camcorder, this requires setting the control switch on the camcorder to the VCR position (but your camcorder may have slightly different settings—you may need to consult the camcorder's manual to see exactly what you must do at this point).

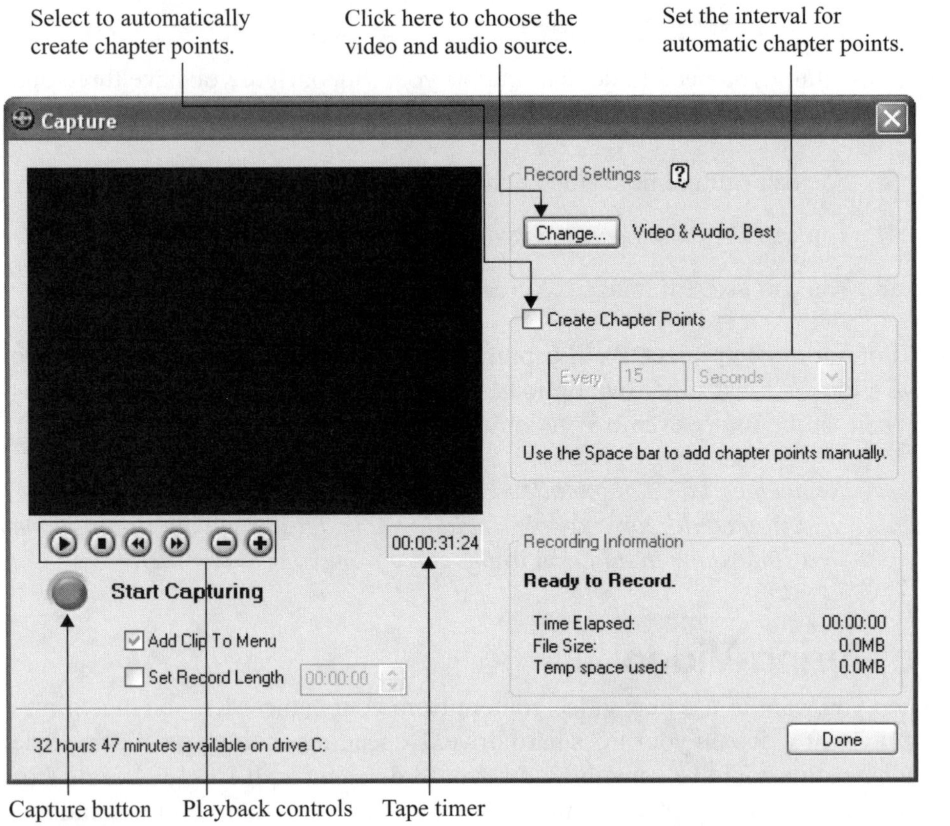

Select to automatically
create chapter points.

Click here to choose the
video and audio source.

Set the interval for
automatic chapter points.

FIGURE 10-2 This dialog box controls video capture in Sonic MyDVD.

Capture button   Playback controls   Tape timer

2. Click the Capture button to open the Capture dialog box shown in Figure 10-2. This same dialog box is used for capturing new video content from analog as well as from digital camcorders.

3. If you want to change any of the settings for the capture (such as the capture device or the quality setting), click the Change button to display the Change Device Settings dialog box shown here. Generally speaking, it's best to choose the highest possible quality settings unless you're trying to squeeze extra content onto the final disc.

Change Device Settings   [X]

Video

Device: Sony DV Camcorder   [v]   [Configure...]

Source: [ ]   [v]   [Source..]

Audio

Device: Avance AC97 Audio   [v]

Line: Mono Mix   [v]

Record Settings

Record: Video & Audio   [v]

Quality: Best   [v]

[ ] Disable preview during video capture

Note: Disabling preview during capture can improve capture quality and performance when using certain capture cards. See documentation for details.

[Restore]   [OK]   [Cancel]

4. When you have verified that you have the correct capture settings, click the OK button to return to the Capture dialog box.

5. If you want to automatically create chapter points at specified intervals during the capture, select the Create Chapter Points checkbox. These chapter points can later be linked to menu buttons to enable easier navigation through your movie. As an alternative, you can simply press the SPACEBAR at any point during the capture to create chapter points manually.

6. To automatically add each chapter to the menu, make certain that the Add Clip To Menu option is selected. Note, however, that if you are capturing video clips that you want to include in your movie without linking them to a specific menu button, this option should not be selected while you are capturing those video clips.

7. Optionally, you can use the Set Record Length option to limit the capture to a specified length of time. I generally find that it is easier to get the results that I want by controlling the capture length manually rather than by specifying a recording length. The exception might be if you are recording something such as a TV show using a TV tuner card (since you probably already know how long the show will run).

8. Use the playback controls to control the playback of a digital camcorder. If you are capturing video from an analog source, you'll have to control the playback manually using the controls on the playback device (a remote control comes in handy if you find yourself in this situation).

9. The Start Capturing button is a toggle. Click it once to begin capturing video at the current position, click it again to stop the capture. You can start and stop the capture as many times as necessary, so that you capture only those portions of the video that you really want to use. You'll need to supply a name for each clip, so if you're capturing from an analog source, be sure to pause the playback while you save the file.

10. When you have finished capturing your video, click the Done button to close the Capture dialog box. In this screen, the scene "Wolf4" is being added to the movie:

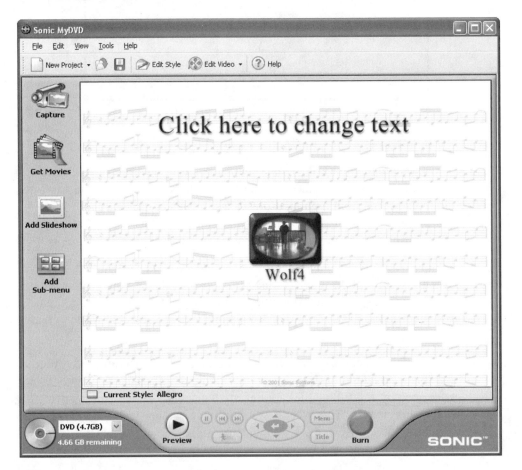

Using Existing Video

Each video clip you capture is automatically saved on your hard drive. When you're creating a movie, you may find that you want to add some video that you've saved earlier in addition to new content that you've just captured. For example, you might have several tapes from your latest vacation, and you might have captured that video on a couple of different days rather than all at once. When you begin creating your movie, you might decide to add some of that existing footage to your movie.

TIP	*You can also use the method discussed in this section to add movie files from someone else to your movie. This makes it easier to create a movie in collaboration with another person.*

To add existing video clips to your movie, follow these steps:

1. Click the Get Movies icon (along the left side of the main Sonic MyDVD screen) to open the Add movie(s) to menu dialog box, shown here.

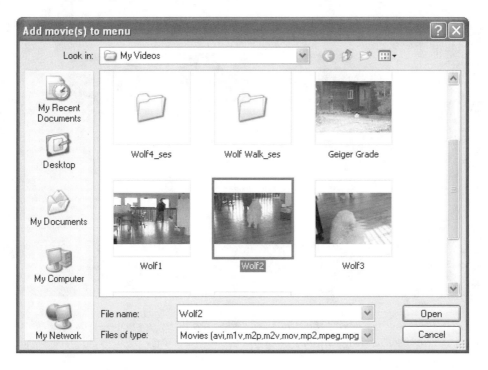

2. If necessary, navigate to the folder containing the videos you want to add.

3. Select the videos you want to add. You can select a contiguous set of videos by clicking the first one, holding down SHIFT, and clicking the last one in the set. You can select multiple noncontiguous videos by holding down CTRL as you click each video you want to include.

4. Click the Open button to add the video clips to your movie and to close the dialog box. As this shows, the selected videos are added just as if you had used the Capture dialog box to capture and add them to your movie.

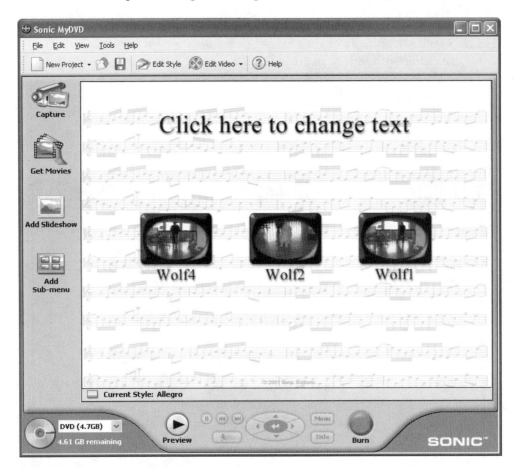

TIP *Don't worry about the names of the video clips for now—you'll get the chance to change them later.*

Adding Still Images

Finally, let's have a look at adding some still images to your movie. You might, for example, want to add a scanned photograph of your grandparents' wedding to a movie about their 50th anniversary. Or perhaps you want to blend some vacation pictures from your digital camera into your movie about your trip to the mountains. Whatever the reason, it's very easy to add those images. You can even create an automatic slideshow complete with background music.

To add some still images to your movie, follow these steps:

1. Click the Add Slideshow icon (along the left side of the main Sonic MyDVD screen) to open the Create Slideshow window.

2. Click the Get Pictures button to display the Open dialog box, as shown here.

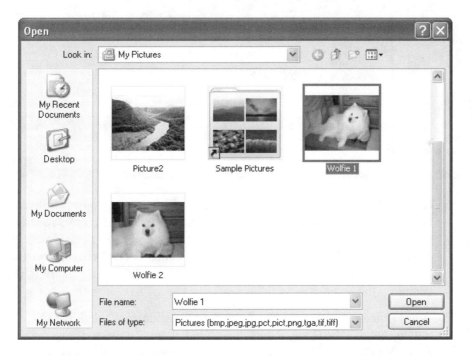

3. If necessary, navigate to the location of the images you wish to add.

4. Select the images (remember that you can select more than one using the SHIFT and CTRL keys) you want to add.

5. Click the Open button to add the images to the Create Slideshow window (as shown here) and to close the dialog box.

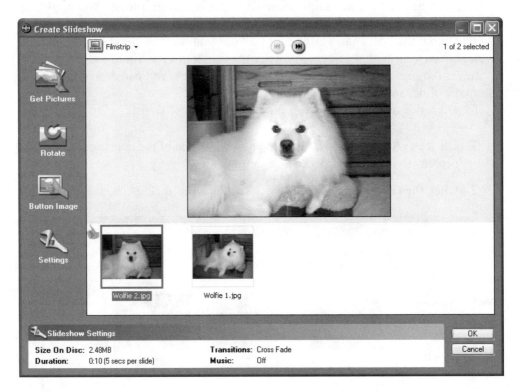

6. If you want to modify the default settings (which are shown in the Slideshow Settings area near the bottom of the window), click the Settings icon. This will display the Slideshow Settings dialog box, where you can choose the duration of the slideshow, background music, transitions between the slides, and the color of the box around the images. Click OK when you are finished changing the settings.

7. When you are finished adding images to the slideshow, click the OK button to close the Create Slideshow window and return to the main Sonic MyDVD window. Once again, your images will be added to the movie—but this time, all of the still images you just added will be a part of a slideshow as shown here:

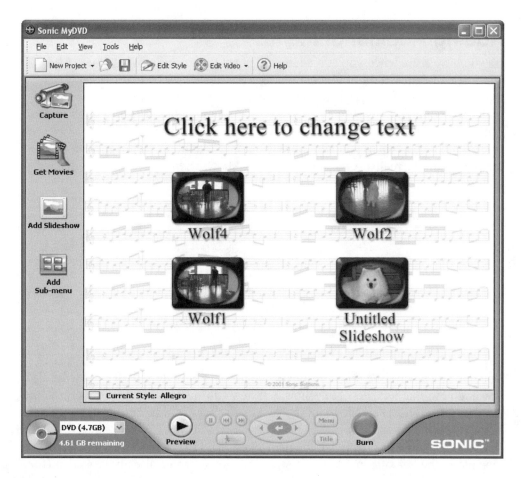

> | TIP | *If you have finished adding videos and images to your movie, take the time*
> *to use the File/Save command before you continue. Doing this will protect*
> *you from losing any of your work in the event that you encounter a problem.*
> *It's a good idea to remember to use this command often as you work—it*
> *takes little time and can prevent so much frustration. I get in the habit of*
> *using the CTRL-S keyboard shortcut for this command since that is even*
> *faster than using the menus.*

Modifying Your Menus

Sonic MyDVD does a lot of things for you automatically, but that doesn't mean
that you're stuck with everything just the way it is initially laid out. There are
a number of modifications that you can make to personalize your movies. We'll
look at some of the changes you can make next.

Adding Additional Menus

As you have seen, when you add additional chapters to your movies, Sonic MyDVD automatically adds additional menu buttons for those new chapters. Eventually, though, you run out of room for more buttons on a single menu. Once you've included six video clips, MyDVD creates additional menus for subsequent video. This maximizes the legibility of the buttons on each menu.

What happens when you add more chapters (or another menu) to your movie? It's pretty simple, really. As this alert message shows, Sonic MyDVD automatically adds another menu page to hold the new buttons:

You may not always see this message (since it is so easily turned off), but whenever you add new chapters, new slideshows, or new sub-menus that would increase the number of buttons on a page to more than six, the page will be automatically added to the menu.

You've probably noticed that the main Sonic MyDVD window includes another icon along the left side in addition to the three icons we've already looked at (for adding content to the movie). This fourth icon is labeled "Add Sub-menu," and you can use it to add a new, blank menu page to your movie. Since extra menu pages are automatically added as necessary, this feature might seem pretty unnecessary to you at first. After all, why would you want to add a menu page manually?

To be totally honest, the option of adding menu pages (sub-menus) manually is there to enable you to control the organization of your movie's navigation. That is, by using sub-menus you can group the chapters in a manner that seems logical and appealing to you rather than simply adding them in a totally automated manner. For example, you might want your main menu to link to a series of sub-menus, each of which contains related items. You might have a sub-menu that links to the chapters of your recent vacation movie, another that links to a couple of slide shows of digital images from your family gathering, and yet another that links to some clips about your favorite hobby. In that way, the main menu would simply be a starting point where viewers would first choose the type of content they wanted to view and then would be able to choose specific items to watch.

If you are going to add sub-menus manually, it's important for you to plan ahead. For example, here are some things to keep in mind:

- Remember that any new chapters, slideshows, or menu buttons are always added to the menu that is currently open. If you want to use the main menu as a jumping off point where viewers choose other menus that contain the actual content, begin your project by adding sub-menus to the main menu. Then, use the Capture, Get Movies, or Add Slideshow icons to add the content to the sub-menus, rather than to the main menu.

- Items are added to menus in the same order you add them using the Capture, Get Movies, or Add Slideshow icons. If you want the items to appear in a specific order on the menus, be sure to add them in that order. You can use the drag-and-drop method to move the buttons around on a specific menu page after they've been added, but it's much more complicated to move things between different menu pages (you can't use drag-and-drop because you can see only one menu page at a time).

- Menu pages in Sonic MyDVD are automatically applied for you in a specific position for maximum visibility. Although you can use drag-and-drop to change the order of the buttons on a menu page, the basic page layout is fixed. When you drag-and-drop buttons the menus reformat themselves as soon as you drop a button. For example, if you drag a button from the bottom row to the top row of a six-button menu, the right-most button on the top row drops down to the second row to make room for the new arrival.

For the purposes of our sample movie project, adding sub-menus manually is unnecessary. You're welcome to experiment with them yourself, and I'm quite certain that you'll have no trouble adding and using them now that you understand how they work. For now, though, let's move on to other parts of building our movie.

Selecting a Menu Style

Since the menus are the first thing that anyone watching your movies will see, you'll probably want to play around with the menu style options to get exactly the look that suits your movie. To do so, you use the Edit Style dialog box shown in Figure 10-3.

Choose the style type.   Select animated menu options.   Play all option
                                View your changes here.   Select the text attributes.

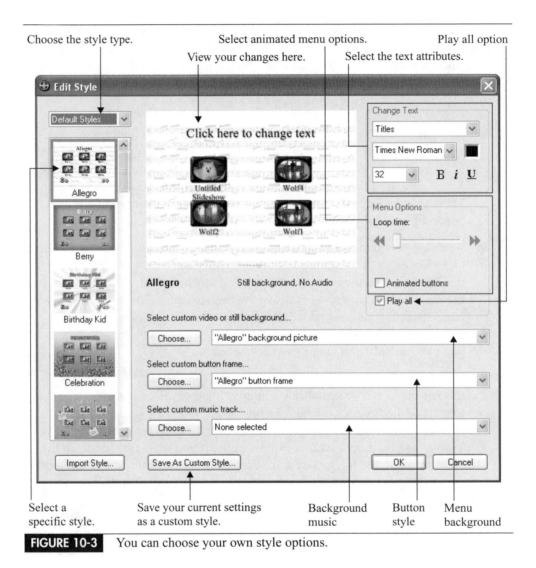

Select a   Save your current settings   Background   Button   Menu
specific style.   as a custom style.   music   style   background

FIGURE 10-3   You can choose your own style options.

Let's take a look at modifying the menu style in closer detail. To begin, follow these steps:

1. In the main Sonic MyDVD window, click Edit Style to open the Edit Style dialog box (see Figure 10-3).

2. Use the drop-down list box in the upper-left corner of the dialog box to choose the style type you want to use. You can choose either Default Styles or Default

Motion Styles from this list. When you choose a different option in this list box, the displayed list of styles will change to show the available styles.

3. Next, select a specific style from the list that runs down the left side of the Edit Style dialog box. You can scroll down to see additional choices. As you make selections, the upper-middle section of the dialog box will automatically change to show your current selections.

4. Use the options in the Change Text section (at the upper-right of the dialog box) to modify the attributes of the text in your menus. You can choose to change the menu title and button text separately or to apply the changes to both at the same time using the list box at the top of this section. It's probably going to look best if you choose the same typeface for the title and the buttons—keep in mind that relatively simple typefaces are easier to read, especially when your movie is viewed on a TV. It's also important to keep the text size fairly large (again, to improve readability).

5. If you want to have movie clips (rather than still image thumbnails) on the menu buttons, use the first two options in the Menu Options section (just below the Change Text section). The Loop time option controls how much of the beginning of each clip is played on the button. The Animated buttons option toggles between still images and playing movie clips on the face of the buttons.

6. Select the Play all checkbox if you want the chapters to play through from the selected chapter to the end of the movie. Deselect this option if you want the menu to reappear after each chapter finishes playing. This selection applies to all of the video clips.

7. To change the background video or image that appears behind your menus, use the menu background selections. You can use the Choose button to browse for more options using the Open dialog box.

8. Use the button style selections if you want to choose a different type of button. You can click the Choose button to browse for more options using the Custom Buttons dialog box.

9. If you want to have music playing while your menus are displayed, use the background music options to select the music. The associated Choose button enables you to select a music file from your hard disk.

10. Once you have made your selections, you may want to click the Save As Custom Style button to save those settings as a named style. Then, you'll be able to quickly choose the same set of options in the future by making a selection from the list along the left side of the dialog box.

11. Click the OK button to close the Edit Style dialog box and return to the main Sonic MyDVD window. This will also apply any changes you've made in the dialog box. For example, here I've selected a different menu style and have increased the size of the title text somewhat:

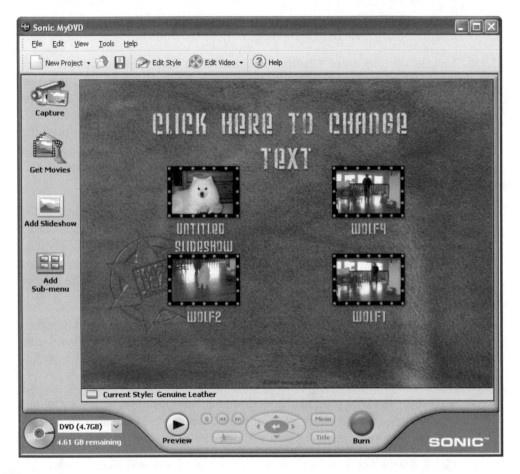

TIP	*Sonic provides a style swap for registered MyDVD users. You can go to styles.mydvd.com and download additional styles for free. These additional styles can be imported into your copy of MyDVD through the same Edit Styles dialog box. You can also purchase additional professionally created styles at Sonic's store, estore.sonic.com.*

Modifying the Text

Let's face it—a movie title like the default "Click here to change text" probably isn't going to impress too many people. It's pretty obvious that you'll want to change the default title to something just a bit more personalized. It's also pretty likely that you'll want to change the default names on the menu buttons, too.

Fortunately, it's very easy to modify all of these titles. Here's how to do so:

1. If you intend to change the menu's style, do so before you edit the titles and button text. Different title styles use different typefaces, and this change can affect how much text you can safely add to a title.

2. Next, select View/Show TV Safe Zone from the Sonic MyDVD menu. If you recall the discussion from Chapter 7, you know that it is important to do this to make certain that your titles will be completely visible when your movie is viewed on a TV set.

3. Click the title to select it, as shown here:

4. While the title text is highlighted, simply enter your new text. If you have somehow managed to select the text box without selecting the whole title, you can use the BACKSPACE key to delete the existing text (or you can click outside the title and then click it again to once again select the text).

5. Click each button's text and modify it. You may want to use button titles as simple as "Chapter 1," "Chapter 2," and so on. Or you may want to add short descriptive titles. Just remember that the text may be somewhat difficult to read on a TV screen, so you should keep it as short and understandable as possible. Here, I've edited the movie title as well as the text of each of the buttons:

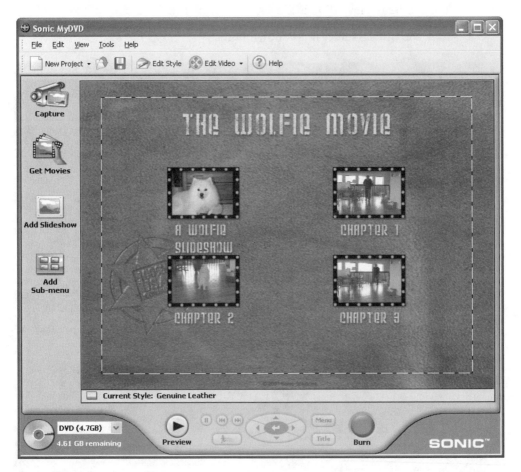

TIP	*Be sure that your title doesn't overlap the buttons. You may need to reduce the font size to prevent this from happening.*

Adding a Soundtrack

Sonic MyDVD Plus offers the basic ability to add soundtracks to your projects. If you want more flexibility to edit your music to match your video, you can upgrade to MyDVD Video Suite. Here are your soundtrack options in Sonic MyDVD Plus (if you do not have ArcSoft ShowBiz):

- You can use the soundtrack that was recorded by your camcorder along with the video. This, of course, means that the movie will include any noise that your camcorder picked up (which may include you giving directions to the people from behind the camera, dogs barking, obnoxious slobs who think everyone should hear the overly loud exhaust sound of their completely unimpressive motorcycles, and so on), so you'll want to remember this when capturing your video.

- You can add music files that you've copied to your PC by dragging and dropping them onto menu buttons in Sonic MyDVD. To modify those music files once they're a part of your movie, you may want to use MyDVD Video Suite.

Fortunately, upgrading to MyDVD Video Suite is pretty simple. Visit the Sonic MyDVD Web site (www.mydvd.com) to learn how you can upgrade your copy to add MyDVD Video Suite at a discounted price. As you'll see in the next section, this is an upgrade that will greatly expand your digital video (and audio) editing toolbox.

Editing Your Video

As you have seen in the earlier sections of this chapter, Sonic MyDVD Plus is designed to be as simple as possible to use and provides basic video editing capabilities. To apply transitions, to add titles, and to include other effects, you'll want to use Sonic MyDVD Video Suite.

In this section, we'll look at how ArcSoft ShowBiz, an add-on program included in the Sonic MyDVD Video Suite version allows you to expand the video editing capabilities of Sonic MyDVD. As you'll see, ArcSoft ShowBiz is a really great addition to the Sonic MyDVD toolbox.

Sending Your Video to ArcSoft ShowBiz

In order to edit your movie, you must first send the portion of the movie that you want to edit from Sonic MyDVD to ArcSoft ShowBiz. The ArcSoft ShowBiz software has been integrated with MyDVD to allow you to move your projects efficiently from one application to the other. You will know when you have switched between the two programs by the difference in the way they look.

Depending on what you would like to accomplish, you can send either a single video clip or all of the video clips in the movie to ArcSoft ShowBiz. You might want to send a single clip if you feel that you need to concentrate on fixing up that one clip. In most cases, however, you'll probably want to be able to edit the movie as a complete entity, so that you can apply changes that affect more than one clip at a time.

To send your movie to ArcSoft ShowBiz for editing, click Edit Video in Sonic MyDVD and choose "Send Selected clip" to edit a single video clip, or choose "Send all clips in menu" to edit a group of clips together. The first of these options will be grayed out and unavailable if none of the clips is currently selected. Regardless of which of the options you select, only video clips—not slideshows—will be sent to ArcSoft ShowBiz.

| NOTE | *If you want to combine several clips into one movie, you can send them all to ArcSoft ShowBiz, and they will be turned into a single clip when they are returned to Sonic MyDVD. However, because you are creating a brand new movie, any chapter points are not carried over.* |

Figure 10-4 shows how ArcSoft ShowBiz appears after you have sent a series of clips from Sonic MyDVD. As you can see, this has a different look from Sonic MyDVD.

Let's take a closer look at some of the important areas of the ArcSoft ShowBiz screen that you'll be using:

■ Near the upper-left of the ArcSoft ShowBiz screen are a series of buttons. These serve the functions of the menus you see in most other programs.

■ In the center of the ArcSoft ShowBiz window is the viewer window where you can see the effects of your editing.

■ Just above the right edge of the viewer window are three buttons that enable you to control the size of the viewer window.

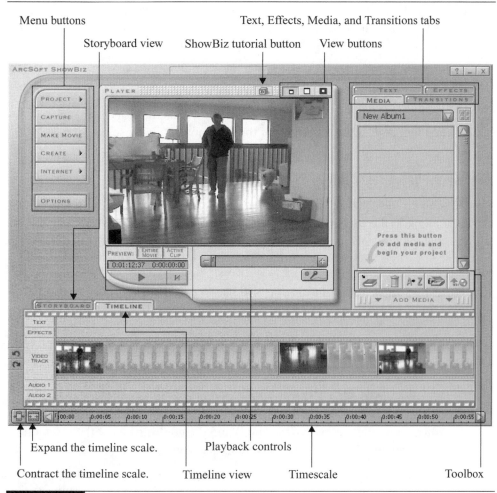

Menu buttons

Text, Effects, Media, and Transitions tabs

Storyboard view   ShowBiz tutorial button   View buttons

Expand the timeline scale.

Contract the timeline scale.

Playback controls

Timeline view   Timescale   Toolbox

FIGURE 10-4   This is the basic ArcSoft ShowBiz screen.

- Just to the left of the view control buttons is the ArcSoft ShowBiz Basics button, which provides tutorial information about various aspects of ArcSoft ShowBiz.

- Just below the viewer window is a set of playback controls. When a single clip is selected, you'll be able to use the controls in this area to edit the selected clip.

- Along the right side of the ArcSoft ShowBiz window is an album with tabs for text, special effects, media clips, and transitions. You use this album to add items to your movie by dragging-and-dropping them onto the timeline.

- Just below the album is a toolbox with several buttons you can click in order to work with items from the album.

- The bottom of the ArcSoft ShowBiz window has the storyboard or the timeline, depending on which of the two tabs you have selected.

Editing Video Clips

Now let's take a look at how you can edit a video clip in ArcSoft ShowBiz. If you've followed along in the earlier chapters, most of this should be pretty familiar to you. It's true that the controls in ArcSoft ShowBiz are slightly different from the ones you've seen earlier, but they are similar enough so that this whole process should go pretty smoothly for you.

Figure 10-5 shows the controls you will use to edit a video clip in ArcSoft ShowBiz. In this case, I've clicked the Enhance button to show the controls for enhancing a video clip.

Now that you've had a look at the controls, let's see how you use them to edit a video clip:

1. Click the video clip (in the timeline) that you want to edit. Notice that the Active Clip button below the left side of the viewer window is activated (as indicated by the yellow background) and the controls below the right side of the viewer window are displayed so that you can edit the clip.

2. To trim the beginning of the clip, play the clip until you reach the point where you want the clip to begin. Then drag the green scissors slider from left to right until it won't move any further.

3. To trim the end of the clip, play the clip until you reach the point where you want the clip to end. Then drag the red scissors slider from right to left until it won't move any further.

4. To use the enhanced controls, click the Enhance button to drop down the Enhanced Controls toolbox.

5. Drag the brightness, contrast, hue, and saturation sliders to the right or to the left to adjust the visual appearance of the clip. Dragging a slider to the right increases the associated value, while dragging it to the left decreases the value. Remember that small changes are probably all that you will want to make in most cases, since large adjustments can quickly make the video seem very unnatural.

6. Drag the speed slider to the right to speed up the playback or to the left to create a slow-motion effect. Be sure to select the Smooth checkbox if you slow down the video so that the frame to frame transitions will be smoother.

Drag to trim the beginning of the clip.   Drag to trim the end of the clip.   Set brightness.

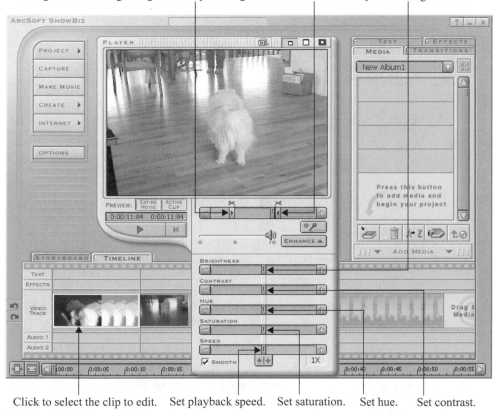

Click to select the clip to edit.   Set playback speed.   Set saturation.   Set hue.   Set contrast.

FIGURE 10-5   You use these controls to edit a video clip.

Modifying the Soundtrack

Next, we'll add some background music to our movie. In this case, I think you'll find that I've used a piece of music that will add a humorous touch.

Here's how to proceed:

1. Click the "red book" button in the toolbox below the album to open the Open dialog box, shown here. (Officially, this is the Add button, but I didn't want you to confuse it with the Add Media button, which appears just below it.)

2. If necessary, navigate to the folder that contains the file you want to add as your background music track.

3. You may want to limit the search (as I've done in the illustration) by making a selection in the Files of type drop-down list box. In this case, I've decided to use an MP3 file as my background music.

4. When you have made your selection, click the Open button to add the file to the ArcSoft ShowBiz media album and to close the dialog box. (I wonder if my dog Wolfie would like my selection?)

5. Click the Add Media button to add the selected music file to the timeline, as shown here (notice that the Audio 1 track in the timeline now shows a waveform to indicate that the track now contains audio).

6. If you want to trim the beginning of the audio track, play the track until you reach the desired starting point, then drag the green scissors slider all the way to the right until it won't move further.

7. If you want to trim the ending of the audio track, play the track until you reach the desired ending point, then drag the red scissors slider all the way to the left until it won't move further.

8. To make the audio track fade in or fade out, select the Fade In or the Fade Out checkboxes (as appropriate).

9. Finally, to view the movie with the soundtrack playing, click the Entire Movie button and then use the playback controls.

You can have two audio tracks in your movie. This can be pretty handy if you want to include both background music and narration, for example. You can add a recorded narration track by clicking the record button (the button with the red dot and the microphone below the right side of the viewer window). When you click this button, you will open the Record Audio dialog box shown here:

As you can probably guess without too much trouble, you click the Record button in the Record Audio dialog box to begin recording. You'll probably want to give the narration file a descriptive name so that it is easier to identify once it has been added to the ArcSoft ShowBiz album. You will need to have your microphone connected to the mic input on your PC before you can begin recording.

Adding Transitions

If you recall the discussion of transitions from Chapter 5, you know that transitions can add a lot of interest to your movies. They are especially handy when you want to call attention to a change of scenes, but you can find many different reasons to add them to your movies. ArcSoft ShowBiz offers an interesting range of transitions that you can use.

To add some transitions to your movies, follow these steps:

1. Click the Transitions tab above the ArcSoft ShowBiz album to display the Transitions album.

2. Use the drop-down list box at the top of the album to locate the category of transition that you want to use.

3. Select the transition you want and then drag-and-drop it where you want it to appear on the timeline. Here, I'm adding the Slither transition between the first two scenes.

4. Click the Entire Movie button and the use the playback controls to view the transition effect. You may need to try out several different transitions until you find the one that best suits your needs.

TIP	*If you don't care for a transition that you have added to your movie, right-click the transition on the timeline and choose Delete from the pop-up menu to remove it. You can also use this same pop-up menu to apply the same transition (or a random one) between each of the clips in your movie.*

Adding Titles

In Chapter 7 we discussed the process of adding titles to your movies. Now we'll look at how you can add some titles in ArcSoft ShowBiz. To do so, you use the Text album shown in Figure 10-6.

To add a title to your movie, follow these steps:

1. Click the Text tab to open the Text album.

2. Use the drop-down list box at the top of the Text album to choose the category of text you want to add.

3. Select the type of text from the album.

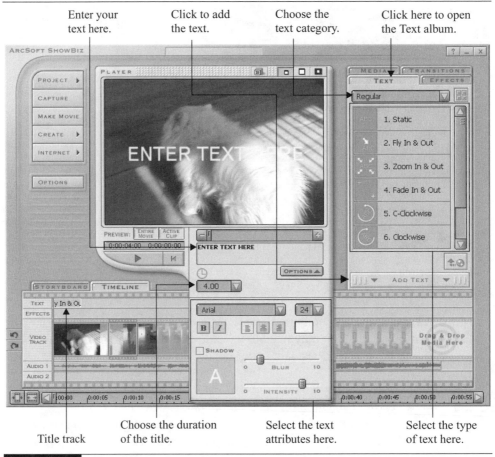

Enter your text here. | Click to add the text. | Choose the text category. | Click here to open the Text album.

Title track | Choose the duration of the title. | Select the text attributes here. | Select the type of text here.

FIGURE 10-6   Use these tools to add text to your movies.

4. Click the Add Text button to add a title track to the text line in the timeline.

5. Type over the default text in the text box just below the viewer window to replace that text with your text.

6. To change the duration of the title, make a selection from the duration box.

7. To modify the text attributes, use the controls in the text options toolbox. If necessary, display this toolbox by clicking the Options button.

8. To move the title to a different location on the timeline, drag-and-drop the title track to the desired point on the timeline.

> TIP
>
> *Remember that the default duration of a title is just 4 seconds, including any motion effects that may be applied to the title. Make certain that the title is displayed long enough so that someone who is not expecting the title will have enough time to read it.*

Adding Special Effects

Finally, we'll take a quick look at the Effects album in ArcSoft ShowBiz. Here I've clicked the Effects tab and selected Filters from the drop-down list box at the top of the album.

It's pretty hard to show the effects of applying one of these filters in a small screen image, so I'll leave it to you to try out some of the items in the Effects album yourself.

Applying one of the effects is pretty simple. You open the Effects album, select the effect you want to apply, click the clip in the timeline where you want to apply that effect, and then click the Add Effect button. Needless to say, you can produce some pretty bizarre results with some of these items (try the Green filter on a video of one of your in-laws, for example).

Returning Your Video to Sonic MyDVD

Once you have finished your editing tasks in ArcSoft ShowBiz, you'll want to return your movie to Sonic MyDVD for completion. This is a fairly simple task, but the procedure may not be completely obvious at first. Let's take a look at how you go about returning your movie to Sonic MyDVD.

1. Click the Project button and choose Save from the pop-up menu. If this is the first time you've saved this movie project in ArcSoft ShowBiz, you'll need to supply a name for your movie project.

2. Click the Create button and choose MyDVD Project from the pop-up menu to display the Export to MyDVD Project dialog box shown here:

3. Enter a name for your movie (and a location if you like).

4. Click OK to begin the rendering process. As this shows, ArcSoft ShowBiz will report on its progress as it is rendering the movie (see the progress percentage indicator in the upper-right corner of the viewer window).

When ArcSoft ShowBiz finishes producing your movie, you will be returned to Sonic MyDVD. The edited version of your movie will now appear on the menu along with the original, unedited versions of the clips. You'll probably want to delete the unedited clips and keep just the edited version on the menu.

Previewing Your Movie

At this point, you're almost ready to burn your movie onto a disc, but before you do it's a very good idea to preview it to make certain that everything works the way you expected. That way, you won't waste an expensive disc if you discover that something isn't quite correct during playback.

The preview function within Sonic MyDVD works just as one on a standard DVD player does. In fact, the onscreen controls that appear just below the preview window during playback mimic the controls you'll find on a set-top player's remote control. This makes it possible for you to test all of the functions to make certain that your finished DVD will do what you expect it to do.

To preview your movie, click the Preview button. The preview playback controls below the viewer window will then be activated, as shown here:

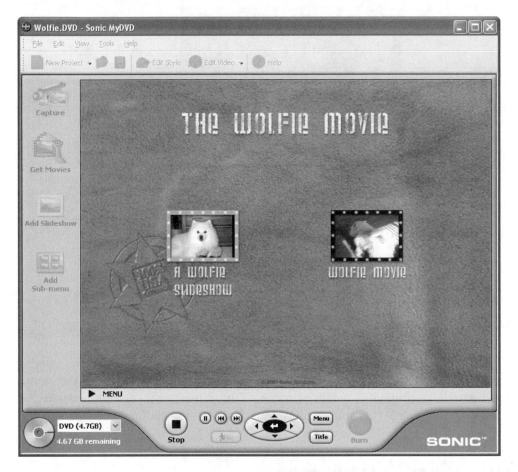

Use these controls to move the highlight, make selections, and control the playback, just as if they were part of a real remote control. When you are satisfied that everything is working properly, you can move on to the final step of burning your disc. Click the Stop button to close the DVD preview mode.

Burning Your DVD or VCD

We're finally down to the one thing you've been waiting for. Now it is time to burn your disc. So, without further ado, let's get on with it!

Make sure you have a blank, recordable disc in your DVD burner (or your CD-R/RW drive if you're creating a VCD). Then follow these steps:

1. Select File/Save from the Sonic MyDVD menu to save your movie project. You can only burn a disc if you have saved the current project and have made no further changes since the last save.

2. Click the red Burn button (near the lower right in the Sonic MyDVD window) to open the Make Disc Setup dialog box shown here:

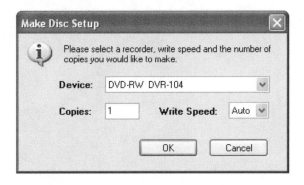

3. Select the drive you want to use (if you have more than one recordable drive).

4. Specify the number of copies you want to produce (it's faster to burn multiple copies all at once, since the movie will have to be rendered only once no matter how many copies you make at the same time).

5. Unless you know that your drive is having problems burning at the highest possible speed, make certain that Auto appears in the Write Speed list box.

6. Click OK to begin the burning process.

7. If you've decided to produce multiple copies of the disc, watch for the prompts to replace the completed disc with a new one.

8. Have fun handing out copies of your new masterpiece!

NOTE   *Sonic MyDVD burns discs using* OpenDVD *technology. This means that you can reopen a movie that you have burned to a DVD if you want to apply further edits. If you saved the original movie on a re-writable disc you can even save your changes to the same disc (but you'll have to use a new, blank disc if you're using write-once discs, of course).*

If you use Sonic MyDVD, this chapter has given you a good start on producing your own movies from start to finish. Along the way, we've stepped through all of the important things you need to know, and I've made sure to show you some tricks to get even more out of the program.

In the next chapter, we're going to repeat this process, but this time we'll be looking at how you can create your movie in Pinnacle Expression.

CHAPTER 11

Creating a Movie in Pinnacle Expression

Pinnacle Expression is one of those well-done computer programs that provide a lot of function while still remaining extremely easy to use. It's hard to imagine that there could be another digital video editor that offers more while still being quite so logical and simple.

In this chapter, we'll create a sample movie from beginning to end in Pinnacle Expression. As you'll see, although the program is simple in appearance, it has some very handy features. For example, once you've captured your video, set up your menus, edited your video clips, and burned your disc, Pinnacle Expression is still not quite done: it also includes the necessary tools for creating professional-looking labels. Now you won't have to hand out discs that have handwritten titles scrawled on them!

One thing you will probably notice immediately about Pinnacle Expression is that there are very few menus. Rather than making you wade through menus looking for what you want, Pinnacle Expression allows you to accomplish almost everything by clicking buttons or icons. While it might seem like this could make the program more difficult to use, in fact the opposite is true. Each of the buttons and icons clearly expresses its function with a logically designed graphic, and each of them also offers a *tool tip*—a pop-up window that appears when you hold the mouse pointer over the button for a few seconds and explains the function of the button or icon.

So, now let's give Pinnacle Expression a try.

Capturing Your Video

The first part of any movie project is always getting the video clips into your computer and ready to use in your video editing program. In this section, we'll have a look at how you do this in Pinnacle Expression.

Figure 11-1 shows the video capture section of the Pinnacle Expression window. To display this section, click the Capture button (the button with the "1" near the top center of the window).

NOTE *Pinnacle Expression does not offer an analog capture option. If you want to use video from an analog source, such as an analog camcorder or a VCR, you'll have to use the analog capture option that was provided with your analog capture hardware to save your video in a file on your hard drive. Once you have done so, you can import that file into Pinnacle Expression using the option to import existing video files (which I'll discuss shortly).*

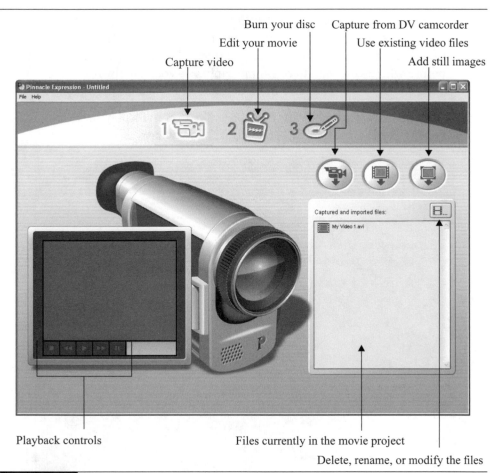

Burn your disc   Capture from DV camcorder
Edit your movie   Use existing video files
Capture video   Add still images

Playback controls   Files currently in the movie project

Delete, rename, or modify the files

FIGURE 11-1   You begin by using these tools to capture your video.

Capturing from a DV Camcorder

Pinnacle Expression can work directly with your DV camcorder to capture your video into a file on your hard drive. When you capture a file in Pinnacle Expression, that file is saved as an AVI file, and it is also imported into your current movie project. Because the captured files are saved as AVI files, you can use those files in later projects as you see fit.

To capture video from your DV camcorder, follow these steps:

1. Make certain that your DV camcorder is properly connected to your PC, that it is powered on, and that it is set for playback.

2. Use the playback controls in the Pinnacle Expression window to position the tape where you want to begin capturing your video. If you're having trouble getting the position exactly correct, place the tape a few seconds before where you want to begin the capture—you can always trim off the extra footage after you have the video clip in Pinnacle Expression.

3. Click the "Capture from DV camcorder" button to display the "Capture video from DV camcorder" dialog box, shown here:

4. Enter a name for the file in the "Enter a name for this capture" text box at the top of the dialog box. Pinnacle Expression automatically supplies a default name of My Video and an incrementing number, but you'll probably want to use something more descriptive, so that you will have an easier time identifying the file in the future.

5. Optionally, you can specify an alternative location to store the captured video. (You can click the folder icon to browse your system if you like.)

6. Click the green Start Capture button to begin capturing your video. Doing this will automatically start the playback on your DV camcorder. Each new scene will be added to the sample menu, as shown here:

7. Click the red Stop Capture button to stop capturing video. If you play the tape all the way to the end of the recorded video, your DV camcorder will stop automatically.

Using Existing Video Files

As you accumulate an ever-larger collection of captured video files, you'll probably discover that some of your existing files would make a great addition to some of your movies. Pinnacle Expression enables you to import any existing AVI or MPEG files and use them in your movies.

Importing an existing video file is a very simple process:

1. Click the Import video file button to display the Open dialog box shown here:

2. If necessary, navigate to the folder that contains the file(s) that you want to import.

3. Select the files you want to import. Remember that you can use the standard Windows modifier keys (SHIFT or CTRL) to select more than one file at a time.

4. Click the Open button to import the selected files. Just as with the videos you capture from your DV camcorder, these new files will appear in the list of captured and imported files.

Adding Digital Photos

Finally, you may want to add some of your still photos to your movie as a slideshow. These might be images from your digital camera, or you might want to use some old photographs that you have scanned. Either way, it's easy to add them to your Pinnacle Expression movies.

Here's how you add digital images to your movie:

1. Click the Import a digital photo button to open the Open dialog box (in this case, the dialog box will display image files rather than video files).

2. Navigate to the folder that contains the images you want to import.

3. Select the files and then click the Open button to import them into Pinnacle Expression.

Modifying the Video Files

Just above the right side of the list of captured and imported files is a button that enables you to rename, delete, or modify the scene detection options for your video files. Clicking this button is the same as right-clicking one of the video files (that is, both actions display exactly the same pop-up menu).

I'm sure that you can understand the first two options—rename and delete—without any extra help. But the third item—Scene detect options—might need a bit of explanation. When you click this item, the Options dialog box shown here appears:

The three choices in this dialog box are:

■ **Detect by Time Stamp (DV only)** This option enables Pinnacle Expression to use the time code that is recorded by your DV camcorder to determine when scenes change. This is the most reliable method of detecting scene changes, since there is a break in the time code whenever you stop recording (and this, of course, is when each scene ends).

■ **Detect by Content** This option is considerably less reliable, since Pinnacle Expression has to guess that there is enough difference between frames to indicate that the scene has changed. Unfortunately, this is also the only way to automate scene detection in imported video files, since they do not contain a time code.

■ **No Scene Detection** This option simply brings in the whole video as one scene, and you must manually add scene breaks if you want them in your movie.

When you have finished adding video and image files to your list, you can move on to the next step of creating your movie.

Choosing a Menu Style

In many ways, the selection of a menu style for your movie can be one of the most important choices that you can make. After all, the menu is the first thing that viewers will see once they have inserted the disc into their player. A well-designed menu that fits the movie's subject can make that all-important first impression.

Pinnacle Expression provides a very impressive array of options to ensure that you can get exactly the correct menu for your movie project. There is a wide selection of pre-made menu templates that you can use as is, and there are also many different ways for you to customize the menus. As a result, you can take the easy path when that suits your needs, or you can tinker when that's your desire.

Let's have a look at your menu options in Pinnacle Expression.

Choosing a Template

In most cases, you will probably find that one of the existing menu templates will be just what you need. It is certainly much easier to simply use one of them than to create your own menu.

Figure 11-2 shows the tools that you use to select one of the Pinnacle Expression menu templates. In this case, I've clicked the button that drops down the menu template drawer so that you can see a few more of the templates.

To choose one of the pre-made menu templates for your movie, follow these steps:

1. Click the Edit movie button (the one in the center above the viewer window) to move to the section of the Pinnacle Expression window shown in Figure 11-2.

2. Use the buttons to the left and to the right of the menu template window to scroll through the available templates one at a time.

3. To view several menu templates at the same time, click the button at the bottom of the template window to drop down the menu template drawer. You can then use the scrollbar along the right side of the menu template drawer to quickly scroll through the templates instead of viewing them one at a time.

When you select a menu template, it is shown in the large viewer window in the center of the Pinnacle Expression window. Your video clips are automatically added to the menu so that you can see how your movie will appear.

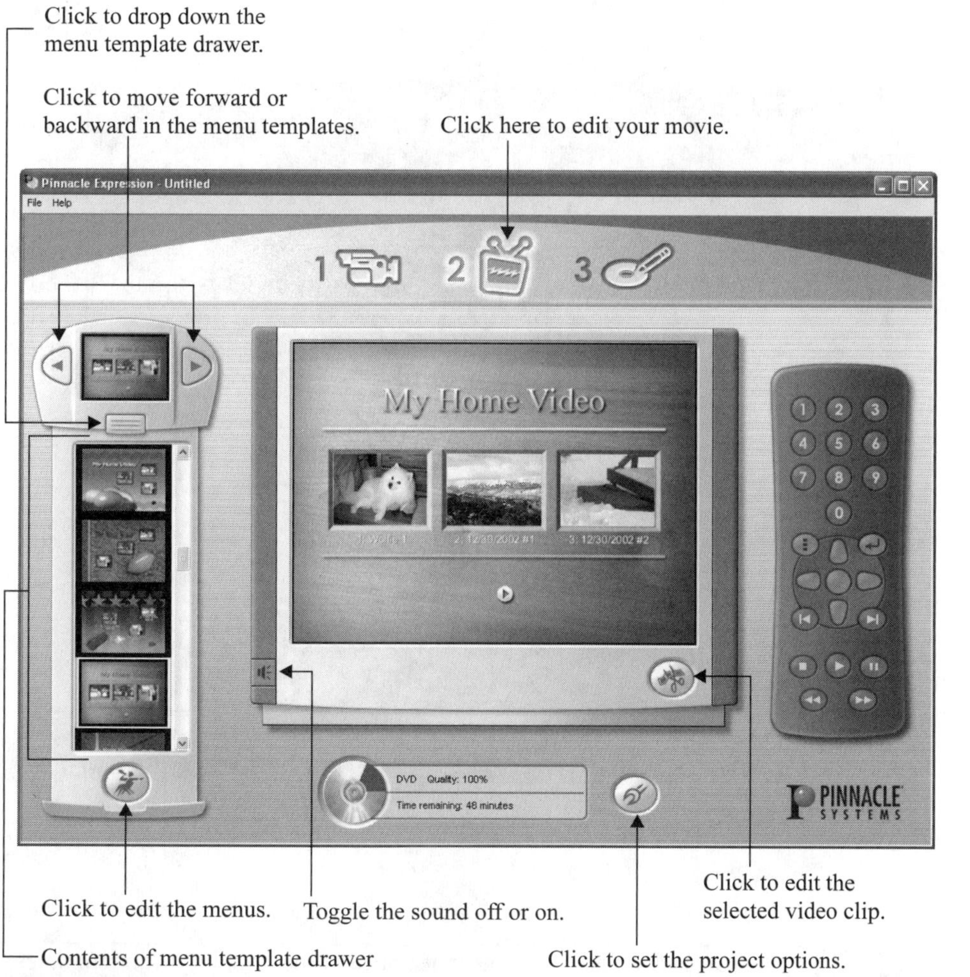
Click to drop down the menu template drawer.
Click to move forward or backward in the menu templates.
Click here to edit your movie.
Pinnacle Expression - Untitled
File Help
My Home Video
DVD Quality: 100%
Time remaining: 48 minutes
PINNACLE SYSTEMS
Click to edit the menus.
Toggle the sound off or on.
Click to edit the selected video clip.
Contents of menu template drawer
Click to set the project options.

FIGURE 11-2 You can begin by using one of the pre-made menu templates.

Customizing Menus

Even though Pinnacle Expression comes with quite a few very good looking menu templates, you aren't limited to the existing selections. You can, in fact, customize any of the templates to get just the look you really want. Figure 11-3 shows the tools that you can use to modify a menu template.

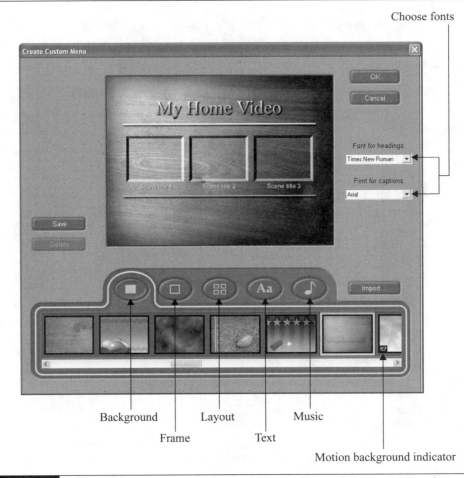

Choose fonts

Background
Frame
Layout
Text
Music

Motion background indicator

FIGURE 11-3 You can easily customize the appearance of the existing menu templates.

Let's take a closer look at how you can customize a menu template. Follow these steps to do so:

1. With the menu template drawer dropped down, click the button at the bottom of the drawer to open the Create Custom Menu dialog box, shown in Figure 11-3.

2. Click the left-most button in the set of five buttons below the viewer window to choose a background image. You can scroll through the list and then click the image that you want to use. Notice the symbol in the lower-left

corner of some of the background image thumbnails indicating that the background will be a video rather than a still image.

3. Click the second button from the left to choose a frame style for the video clip thumbnails that appear on your menus.

4. Click the third button to choose a layout for the buttons on your menu. Keep in mind that some layouts work better with certain backgrounds than others, so you may have to choose a different background image for some layout styles.

5. Click the fourth button to choose a text style for the menu title.

6. Click the fifth button to choose background music for your menu.

7. You can use the drop-down font list boxes to choose typefaces for both the menu title and for the button captions.

8. Once you have made your changes, click the Save button so that you can save your modified menu template for future use.

9. Click OK when you are finished to return to the main Pinnacle Expression window and to apply your selections to your movie.

NOTE	*If you add background music to a menu, Pinnacle Expression sets the duration of the menu between 40 and 280 seconds. If your music clip is within this range, the length of the music clip determines the menu duration. If your music clip is less than 40 seconds, Pinnacle Expression adds enough silence so that the total length of a loop is 40 seconds. If your music clip is longer than 280 seconds, anything beyond the 280 limit is simply cut off, and the music begins playing from the beginning after 280 seconds. If you have a motion background rather than a still image, the length of the motion background video clip determines the menu duration, and this can cause the audio to loop sooner if the video clip is shorter than the audio clip.*

Adding Titles

One additional thing you'll probably want to do is to change the titles of the menus you add to your movies. The "My Home Video" title may be okay in some cases, but to really personalize your movies you'll want to add something that's a bit more descriptive (and not quite so generic).

Changing the title text in a Pinnacle Expression movie is quite easy, but the method that you use might not seem completely obvious to you at first. You might expect to be able to change titles when you customize the menu template (in the previous section), but that's not the case—when you think about it, though, you wouldn't want to modify the template in that way because the template can be the basis for the menus in more than one movie.

Actually, all that you need to do to change the menu titles for your movies is to click the title and type in your own text, as shown here:

In this case, I've replaced the default text with "Sierra Winter" to reflect the theme of my movie. You can also click any of the chapter titles and enter a new name in this same manner to change the chapter's title.

> **NOTE** *Although you can change a chapter's title by simply clicking the title and entering a new title, the chapter number that precedes the title is not changed. Maintaining sequential chapter numbers allows them to be used to navigate your movie. In the next section I'll discuss the procedures for rearranging the chapters so that different video clips are assigned to specific chapter numbers.*

Editing Clips

At this point, you have a movie that contains a number of video clips, possibly a slideshow, a nice-looking menu, and titles that fit your subject. What more could you possibly want? Well, there are these little matters of editing your movie so that the scenes play in the proper order, trimming out unwanted footage, and taking control over the playback options.

In this section, we'll examine the editing options that are available to you in Pinnacle Expression. To access these options, click the Edit button just below the right side of the viewing window in the main Pinnacle Expression window. When you click this button, your movie will appear in the Edit Video dialog box, as shown in Figure 11-4.

Let's take a look at how you can use the Edit Video dialog box to edit your movie.

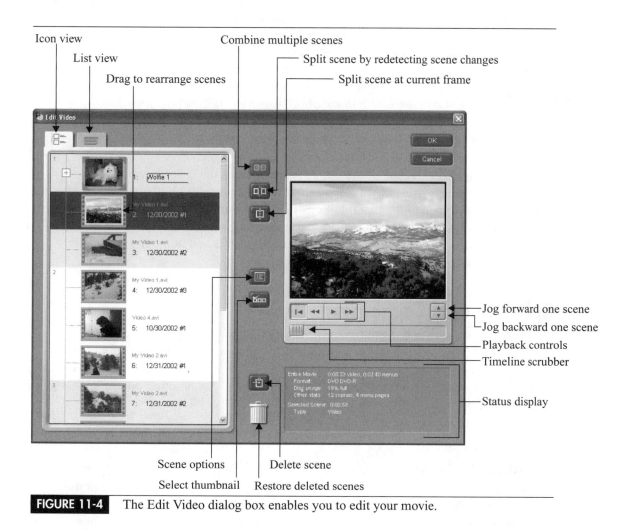

FIGURE 11-4 The Edit Video dialog box enables you to edit your movie.

Rearranging Chapters

As you have no doubt noticed already, when you capture or import video clips, they may not appear in your movie in the best possible order. A little bit of rearranging might be just what your movie needs in order for it to flow more smoothly or to do a better job of telling the story you're trying to tell.

You can easily rearrange the clips in your movie by dragging-and-dropping them within the Edit Video dialog box's list. Simply move a clip upwards to make it appear earlier in the movie, or move it downwards to make it appear later in the movie. When you drag-and-drop a clip, the remaining clips will automatically move as necessary to make room for the clip you're moving (and to fill in the previous location so that there are no gaps in the movie).

> **NOTE** *The Edit Video dialog box offers an icon view and a list view of your movie. You can use either one, but the icon view is generally somewhat easier to use, since it shows a thumbnail view of each clip that allows you to quickly identify the clip you are modifying.*

Rearranging the sequence of the clips also rearranges the order in which those clips appear on the movie's menu. This also affects the chapter numbers, which are automatically assigned to the menu buttons in a strict numerical order. Therefore, if you move a clip up from the 4th position to the 2nd position, that clip will now be assigned to button number 2, while the previous number 2 and number 3 clips will change to number 3 and number 4, respectively.

Splitting and Combining Clips

One of the most basic and useful ways of editing a movie is by splitting clips into multiple clips. This process makes it possible for you to cut out unwanted footage or rearrange footage that appears within a single clip. In Pinnacle Expression, this is also the only way you have to trim a scene (since there are no other controls for trimming the beginning or the ending of a scene).

Let's take a look at your options for splitting scenes.

Redetecting Scenes

The first method of splitting scenes is to click the second button from the top in the middle of the Edit Video dialog box. When you do, Pinnacle Expression will display the menu shown here, and you can choose a method for redetecting scene changes:

Detect scenes by shooting time and date
Detect scenes by video content
Split scenes at regular intervals...
Split scenes at edit points

This option is probably most useful with imported video files (since Pinnacle Expression automatically detects the scene changes in recordings it captures from DV camcorders by using the breaks in the time code). Personally, I find this option to be too imprecise to be very useful. If I'm going to split scenes into multiple scenes, I prefer the more precise control afforded by the next option.

Splitting and Trimming Video Clips

The third button from the top in the Edit Video dialog box enables you to split a scene at a particular frame in the movie. Since you can use the playback controls, the scrubber, and the jog buttons to move to the exact frame you want, this method provides very precise control over the point at which the scene is trimmed.

Figure 11-5 shows a very good example of why such precise control is so important. In this case, I accidentally shook the camera at the beginning of the scene, so the first several frames of the scene are quite blurry.

Because I wanted to remove the blurry frames at the beginning of the scene, I advanced the scrubber to the exact frame where the video became stable, then clicked the button to split the scene at the current frame. The jog buttons under the right edge of the viewer window were very handy in locating the exact frame that I wanted. As shown in Figure 11-6, when I split the scene, it became two consecutive clips in the scene list, and the two clips were both highlighted.

Since I wanted to remove the blurry footage from my movie, I next clicked just the first of the two clips that resulted from splitting the scene. This selected the blurry clip and deselected the clip I wanted to keep. Finally, I clicked the next-to-the-bottom button in the middle of the Edit Video window to delete the blurry clip. This eliminated the footage that I no longer wanted and improved the appearance of my movie.

> **NOTE** *If you split a clip and then delete one of the resulting sections of the original clip, you'll also be deleting any recorded audio that was contained in the deleted clip. Pinnacle Expression does not allow you to edit the video and audio portions of a scene separately, so in some cases you may have to make a compromise between the competing goals of deleting unwanted video footage and retaining as much of the recorded audio as possible.*

FIGURE 11-5 I will want to trim the blurry video from the beginning of this scene.

Recombining Clips

If you split a scene into multiple scenes, each of the new scenes will be attached to its own button on the movie's menu. The extra buttons will disappear if you delete the extra scenes, of course, but if you change your mind and decide that you don't want to delete the new scenes, you may not want the extra clutter on your menu.

You can recombine clips that you have split by selecting the clips and clicking the top button in the middle of the Edit Video dialog box. This button is available only in certain circumstances, however:

- The clips you want to combine must have originally been a part of a single clip.

- The clips must appear in their original order.

- There must not be any footage missing between the clips you want to combine. That is, if you split a clip into three pieces and then delete the center section, you cannot combine the first and third pieces into one clip.

FIGURE 11-6   I have split the scene so that I can delete the poor-quality video.

Deleting Clips

Sometimes you may want to delete scenes that simply have no place in your movie—regardless of their video or audio quality. For example, my scene list has a scene from an interview, which ended up as scene 5, that is clearly not a part of my Sierra Winter movie. In addition, the slideshow in scene 1 just doesn't fit with the rest of the movie, either, so it can go.

As you saw in the previous section, to delete an unwanted scene you simply select the scene and then click the Delete button (second from the bottom). But anyone can make a mistake, and it's pretty easy to accidentally delete a scene in error. If this happens to you, don't panic, because help is available with the click of a button.

If you discover that you've deleted a scene in error, click the Recycle Bin button (the trash can at the bottom of the middle section of the Edit Video dialog box) to display the Recycle Bin dialog box, as shown in Figure 11-7. Then select the scene (or scenes) that you want to recover and click the Restore button. The accidentally

FIGURE 11-7   The Recycle Bin can help if you accidentally delete a scene.

deleted scene will be restored, and you can get on with your editing as if nothing had happened.

Setting Scene Options

The middle button in the center of the Edit Video dialog box enables you to set certain options that apply to the scenes in your movie. Clicking this button displays the Edit Options dialog box shown here:

Edit Options ☒

For all scenes

☐ Show Title When Playing

☐ Return to menu after playing

☐ Fade to black between scenes

☐ Dissolve between scenes

[OK] [Cancel]

Let's take a look at what these options do:

- **Show title when playing** Selecting this option displays the title on top of each scene as it plays. Since you probably don't want the titles to be displayed constantly, you probably won't have much use for this option.

- **Return to menu after playing** If this option is selected, each chapter in your movie will act as though it were an independent movie. Clicking a menu button will show the single clip, and when the clip finishes playing, the viewer will be returned to the menu. Since Pinnacle Expression provides no means of combining noncontiguous clips, you probably won't find this option too useful either (because each chapter will be limited to the length of a single clip).

- **Fade to black between scenes** Selecting this option causes Pinnacle Expression to add a fade transition between each of the scenes in your movie.

- **Dissolve between scenes** Selecting this option adds a dissolve transition between each of the scenes in your movie. These final two options are mutually exclusive, so selecting one of them deselects the other.

TIP *The scene options apply to all of the scenes in your movie. You cannot apply them to individual scenes.*

If your movie contains a slideshow, you have additional options that apply to the slideshow but not to the video scenes. If you click the Scene Options button when a slideshow is selected, the Edit Options dialog box takes on a different appearance, as shown in Figure 11-8. You can then make selections in the lower section of the dialog box to control the options that are specific to the slideshow.

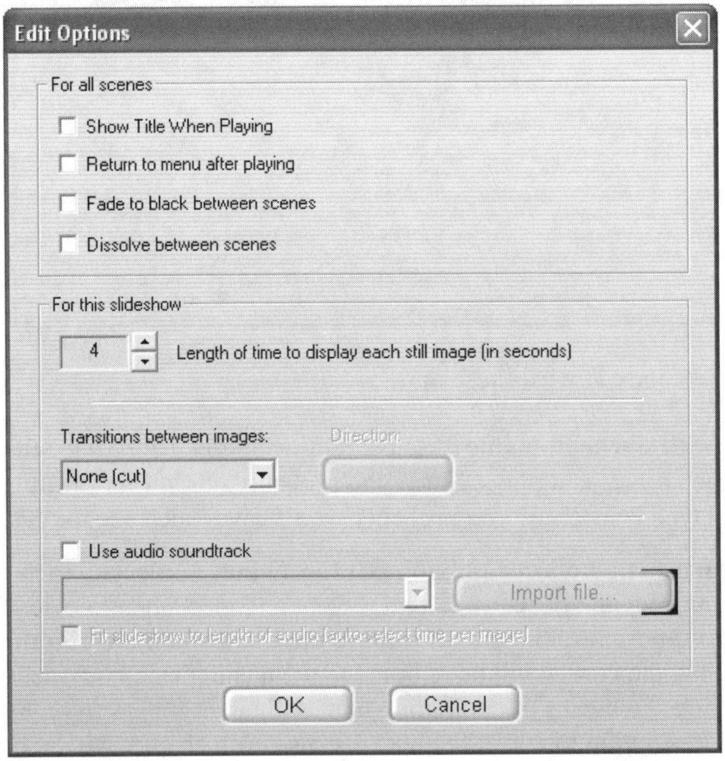

FIGURE 11-8 You have different options when you are working with a slideshow.

Choosing a Different Thumbnail

The *thumbnail* is the image that appears on the menu button that selects the scene for playback. By default, the thumbnail image is always the first frame from the scene. In some cases, however, you may prefer to use a different image for the thumbnail. You can use any of the frames in the scene as the thumbnail.

In Figure 11-9, for example, I've selected scene 2 and then moved the scrubber forward to find an image that might work better as a thumbnail for the scene. Selecting the new image as the thumbnail is as simple as clicking the Select thumbnail button.

TIP *To reset the thumbnail to the first frame of the clip, simply move the scrubber back to the first frame and click the Select thumbnail button again.*

FIGURE 11-9   You can choose any frame you want for the thumbnail.

Previewing Your Movie

When you have finished editing the scenes in your movie, click the OK button to close the Edit Video dialog box and return to the main Pinnacle Expression window. The menu will now reflect any of the changes you have made.

At this point, you should select the File/Save Project command to save your work. If you have not saved this movie project earlier, you'll need to supply a file name for your movie project.

Next, you should preview your movie to make certain that everything is just right. This is an important step before you move on to burning a disc, because you still have the opportunity to correct any problems at this point.

You can play your movie in Pinnacle Expression by using the built-in player. This player works just like a set-top player, so you can test all of the controls by clicking the buttons on the remote control (on the right side of the main Pinnacle Expression window). As shown in Figure 11-10, the viewing window displays your movie as it is being played.

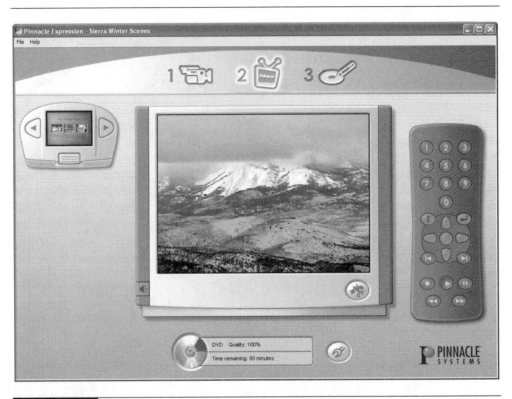

FIGURE 11-10   It is a good idea to preview your movie before burning a disc.

> **TIP**   *To get a better feel for the navigation in your movie, be sure to use the controls on the remote control rather than clicking in the viewer window. Although both methods can be used to play your movie, someone using a set-top DVD player will have to use their remote control to play your movie.*

Burning Your Discs

If you are satisfied with the way your movie plays, it may be time to move on to burning your disc. With Pinnacle Expression you can burn any of the three popular disc formats: VCD (VideoCD), S-VCD, or DVD.

Before you can burn your disc, you need to check the Pinnacle Expression recording settings to make certain that the project is properly set up for the type of disc that you want to create. To do so, click the Settings button to display the Settings dialog box shown in Figure 11-11. In this case, I've clicked the Advanced button

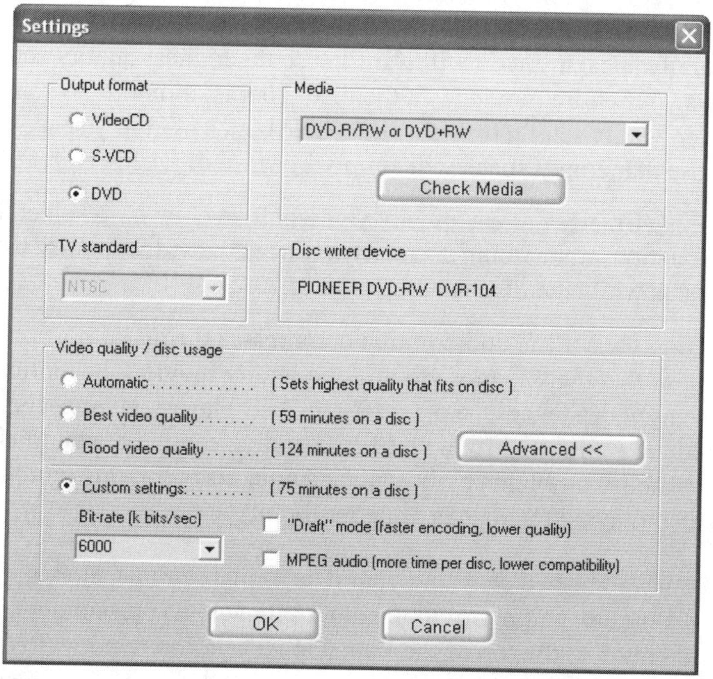

FIGURE 11-11   Make sure that you select the correct output settings for your project.

in the Settings dialog box and selected the Custom settings option so that all of the options in the dialog box will be visible.

The Settings dialog box has a number of important options that you will want to understand:

- The Output format options control the type of disc that you will create. Both of the first two options are used to create a disc that you will burn onto a CD-R blank, while the third option requires a recordable DVD.

- The Media options select the type of blank media that you will use for recording. Your choices here interact with the Output format options—VCDs and S-VCDs must be recorded on a CD-R (or CD-RW), and DVDs must be recorded on recordable DVDs.

- You cannot make a selection in the TV standard section—this setting is advisory only. Your copy of Pinnacle Expression can only make discs that conform to your local TV broadcast standard.

- The Video quality/disc usage section enables you to choose the proper balance between the video quality and disc capacity to suit the needs of your movie project. These settings can be made only for S-VCD and DVD movie projects (VCDs use a standard setting that cannot be changed). Generally speaking, you will want to use the highest quality setting that allows your entire movie to fit the disc. That is, it makes no sense to choose a lower quality setting that leaves a whole bunch of empty space on the disc, because that empty space will simply be wasted.

■ There is virtually no reason why you would ever want to select the Draft mode option, since doing so simply reduces the video quality to save a few moments while the disc is being created.

■ Selecting the MPEG audio option compresses the audio portion of your movie so that there is more room for a longer movie. This option can cause problems with a few of the older set-top DVD players, but most newer players should have no difficulty with MPEG audio. You may want to check for compatibility problems by burning a single disc with this option selected and then playing the disc in your set-top player.

When you have finished choosing your disc format settings, click the OK button to close the dialog box and return to the main Pinnacle Expression window. Next, click the third button at the top of the Pinnacle Expression window to display the disc creation section shown in Figure 11-12.

To burn your disc, follow these steps:

1. Insert a blank disc of the proper type in your recordable drive.

2. If you want to make multiple copies of the same disc, specify the number you want using the Number of copies option. Remember that since the *transcoding* (rendering) of your movie needs to be done only once no matter how many copies you burn at the same time, you can save considerable time by burning all of your copies at once.

3. If necessary, click the Settings button to specify the burning speed and the location of the temporary file directory. In most cases, you can simply leave these options set to their default values.

4. Click the Start button to begin the process. The transcoding progress bar will indicate the progress of the transcoding process, and when that is complete the disc burning progress bar will indicate the progress of that process as well. Don't disturb your system until both of these processes have completely finished—otherwise, you may waste a blank disc.

5. If you are burning more than one copy of your disc, watch for the prompt to insert a new blank when the first copy is finished.

TIP *It's a good idea to make one copy of your disc and test it the first time you burn a disc with Pinnacle Expression. That way you can determine if there are any problems before you create a whole bunch of unusable discs. Once you know that everything is functioning properly, you can go ahead and burn multiple copies.*

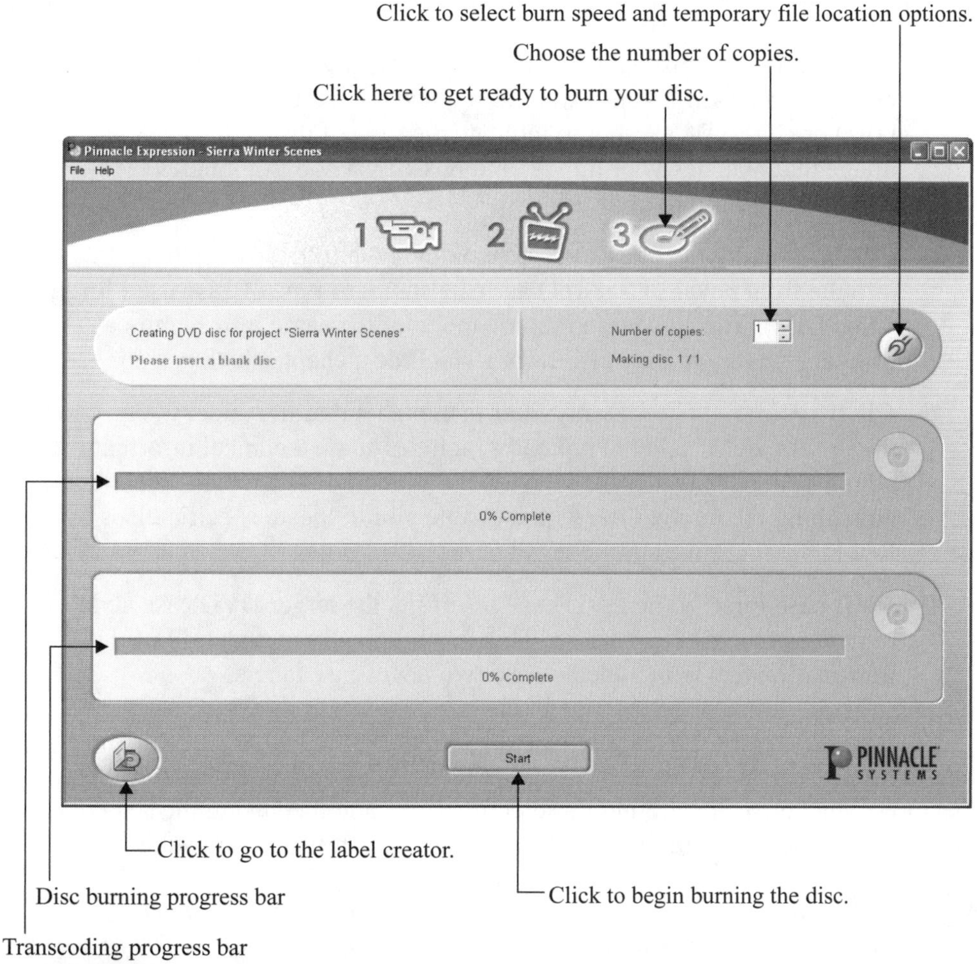

Click to select burn speed and temporary file location options.

Choose the number of copies.

Click here to get ready to burn your disc.

Click to go to the label creator.

Disc burning progress bar

Click to begin burning the disc.

Transcoding progress bar

FIGURE 11-12 This is the disc creation section.

Labeling Your Discs

When you have finished burning your disc, you'll want to complete your project by creating some fancy looking labels. Pinnacle Expression makes this quite easy by including a disc labeling application that can turn out labels for your disc as well as for the jewel case or DVD case.

Figure 11-13 shows the window that appears when you click the Create labels button on the disc burning section of the Pinnacle Expression window.

The disc labeler contains four tabs that you can use to create the different types of labels you might need to complete your project. These label types are the following:

- **Disc label**   The disc label is the label you apply to the disc itself. Generally speaking, you'll want to include the name of the movie, the DVD or CD-ROM symbol to indicate the type of disc, a background image that matches your movie, and possibly a copyright notice. Anything else might be overkill.

- **CD case booklet**   The CD case booklet is the folded-over booklet that fits into the front cover of a jewel case (the standard type of case used for most audio CDs). You'll want to include much of the same information as on the disc label, and you may also want to include a chapter listing.

- **CD tray liner**   The CD tray liner is the label that fits under the tray in the jewel case. This label typically includes some technical information about your disc. This is also the label that includes the small side labels containing the movie title, which enable you to locate specific discs by looking at the spines of the jewel cases.

- **DVD case liner**   The DVD case insert fits the larger DVD case style rather than the CD jewel case. This label typically includes all of the information that is included on the two jewel case labels.

You can edit any of the text on a label by clicking the text to select it and then entering replacement text in the text box at the bottom of the disc labeler. You can also move label elements around by selecting them and then using the selection handles that appear around the object to drag the object where you want it.

> **TIP** *When an object's selection box has 12 selection handles, you can use the outer set of handles to bend the object into different shapes. The inner set of handles resizes the object without bending it. Click the Pinnacle Systems graphic on the disc label tab to try this out for yourself.*

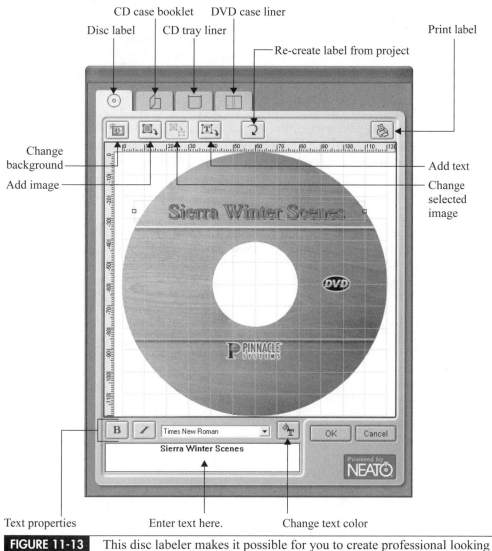

CD case booklet — DVD case liner

Disc label — CD tray liner

— Re-create label from project

Print label

Change background —

Add image —

— Add text

— Change selected image

Text properties

Enter text here.

Change text color

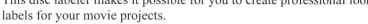

FIGURE 11-13 This disc labeler makes it possible for you to create professional looking labels for your movie projects.

When you have finished creating your labels, you can click the Print button to print the labels immediately, or you can click the OK button to save the labels for later printing. This latter option is handy if you will be making additional copies of your disc in the future, since all of your label settings will be saved with your movie project (just remember to save your movie project before you exit from Pinnacle Expression).

> **TIP**   *Be sure to use high-quality labels. If possible, get the ones that say they are specifically designed for use on DVDs, because the higher rotational speed can easily throw a poorly labeled disc off balance.*

In this chapter, you've learned how to create a complete movie project using Pinnacle Expression. We started out with capturing and importing video content as well as still images for slideshows. Next, we looked at choosing a menu template and saw how you could customize the template to suit your needs. From there, we moved on to editing the video clips to produce the best looking movie possible. Following that, you saw how to create your disc by recording your movie onto it. Finally, you saw how Pinnacle Expression gives you that extra little touch: a disc labeler so that your movies will have that professional appearance you find on discs you buy.

Next, we're going to have a look at the movie creation process using another digital video editing program, Roxio VideoWave Movie Creator.

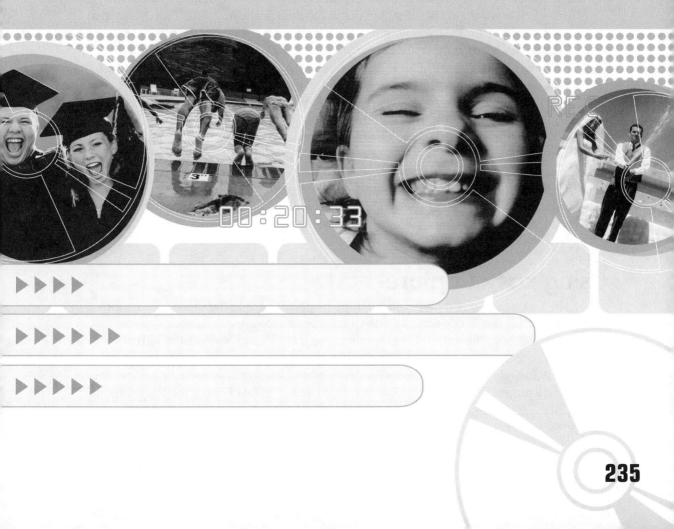

You may not have heard very much about Roxio VideoWave Movie Creator yet, but it's definitely a program you'll want to know about. Even though this program is aimed at users who just want to get a movie made with as little fuss as possible, Roxio VideoWave Movie Creator is loaded with all sorts of features that will please even those users who want something more than basic digital video editing capabilities. Sure, this program is very easy to use, but it's also packed with features you probably wouldn't expect to find.

Before we begin the substance of this chapter, I'd like to point out that Roxio VideoWave Movie Creator really offers you a number of different choices of how to go about creating your movies. These range all the way from a simple template-based option to an editor that gives you considerable control over most of the movie's elements. In a sense, Roxio VideoWave Movie Creator is like several programs in one. When you want to turn out a video in a hurry, you have that simple option. But when you want to create something a little fancier, you've got a lot of choices.

In this chapter, I'll try to hit most of the high points of Roxio VideoWave Movie Creator, but because there are so many different paths you can take through the program, you may not use all of the options when you create your own movies. I think, however, that you'll find that it's awfully nice to be able to make those choices.

Figure 12-1 shows the main Roxio VideoWave Movie Creator window. Throughout this chapter, I'll be asking you to click buttons along the left side of this window, but I won't be showing this main window again. I'm quite certain that you'll easily be able to understand what I'm asking you to do without the window being displayed. Incidentally, be sure to notice how the descriptions in the right- side area of the window change as you move the mouse pointer over different options. These descriptions can provide that little bit of extra help you need in selecting the option you want.

Using Easy Capture

The first step in any digital video editing project is getting the video content into your PC. Roxio VideoWave Movie Creator seems to be especially versatile in accomplishing this task. This program is capable of using video from virtually any content source you might want to throw at it.

Although you can add further video content later, the way to begin almost any movie project in Roxio VideoWave Movie Creator is by clicking the Easy Capture

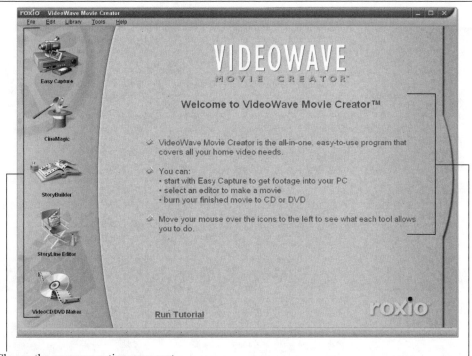

Choose the program option you want
by clicking one of these buttons.　　　　　　　View a description of the selected option here.

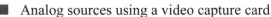

FIGURE 12-1　　The buttons along the left side of the Roxio VideoWave Movie Creator
window provide access to the various parts of the program.

button on the main window. This will display the first step of the Easy Capture
section of the program, as shown in Figure 12-2.

Selecting a Video Source

Roxio VideoWave Movie Creator allows you great latitude in choosing the source
of your video content. Your options are the following:

- DV Camcorder
- Analog sources using a video capture card

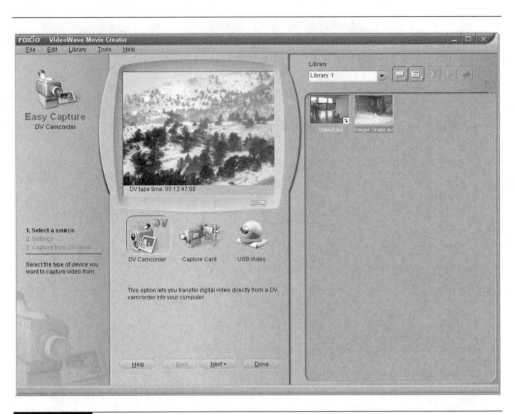

First, choose your video source.

- USB video cameras
- Existing video files in virtually any Windows-compatible video or digital image format

To choose one of the first three options, click the appropriate button below the viewer window. Keep in mind, however, that only those video sources that are actually attached to your PC can be used. You won't, for example, be able to select USB Video unless you have a USB camera attached to your PC. If you select an option that is currently unavailable, the viewer window will go blank.

To open an existing video file, click the Add file(s) to Library button (the button that looks like a filmstrip just to the right of the Library drop-down list box). This will display the Add Files to Library dialog box, shown here, so that you can select the files you want to add.

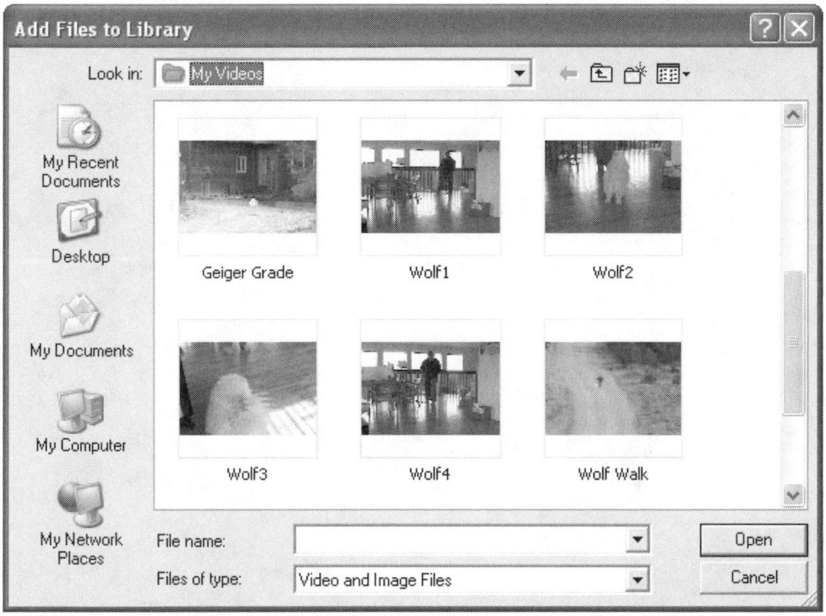

Click the Open button once you have selected the files to add them to the Roxio VideoWave Movie Creator library (the large area along the right side of the Easy Capture section of the window).

Choosing Your Settings

Once you have selected the video capture source, click the Next button to continue. When you do, you'll be given a set of options specific to your selected capture source. For example, if you're capturing video from a DV camcorder, you can choose DV or MPEG-2 for the capture format (MPEG-2 sacrifices a small amount of video quality but offers a much more compact video file since the video is compressed). Similarly, the other two capture source options enable you to make size versus quality selections appropriate to the capture source.

In each case, you will also find a Settings button on the video capture format page. Clicking this button displays a Capture Settings dialog box, where you can choose such options as the folder you want to store the captured video in, the automatic file naming convention to be used, and so on (as shown in Figure 12-3).

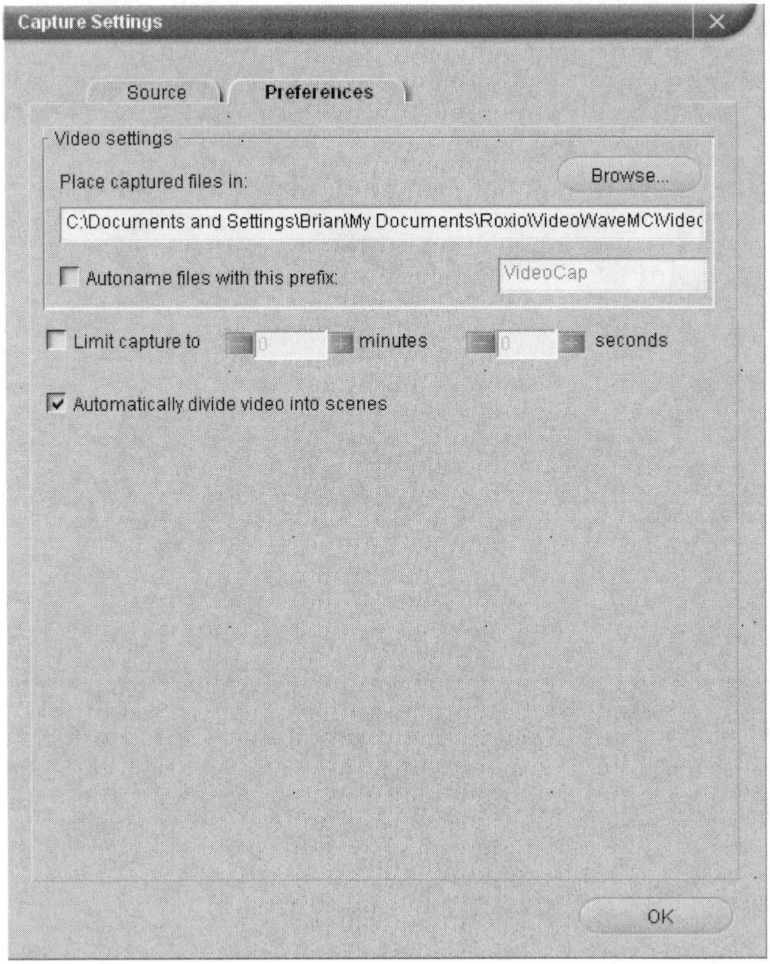

FIGURE 12-3 Select your capture settings.

Note that the "Automatically divide video into scenes" option is available only if you have selected DV Camcorder as the capture source.

Capturing Your Video

When you have verified that the capture settings are correct, click the Next button to move on to the video capture page. Figure 12-4 shows the center part of the page as I'm ready to begin capturing video from a DV camcorder.

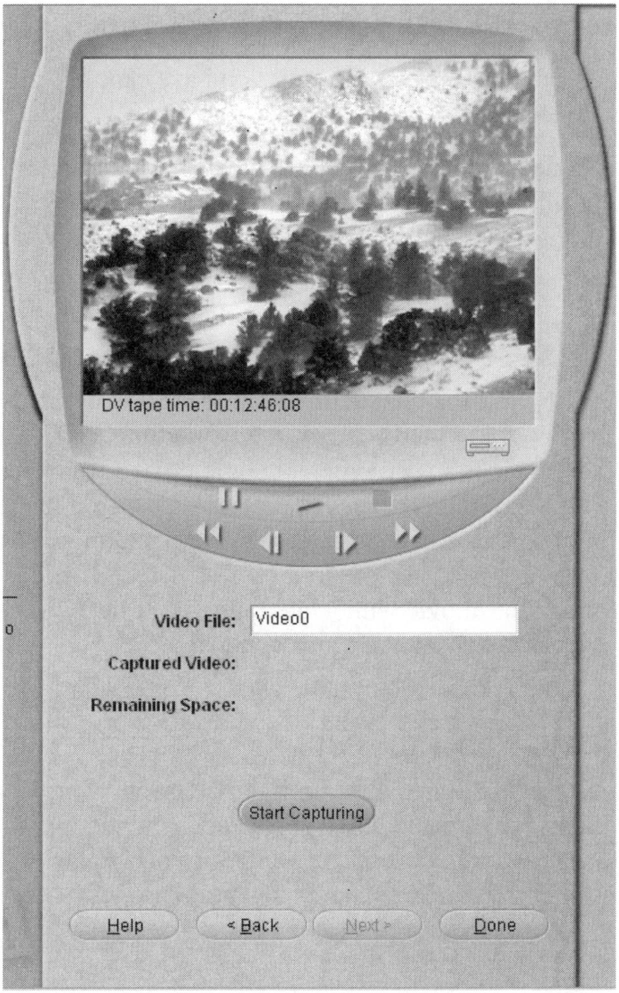

FIGURE 12-4   Capturing video from a DV camcorder

Here's how to capture the video:

1. Use the playback controls below the viewer window (or the playback controls on your analog video source) to begin the playback.

2. When you wish to begin capturing the video, click the Start Capturing button.

3. Click the Stop Capturing button to stop the capture. Your new video will then be added to the Roxio VideoWave Movie Creator library.

4. Repeat the process as necessary until you have captured all of the scenes you want to include in your movie.

5. Click the Done button when you have finished capturing (or importing) your video. This will return you to the main Roxio VideoWave Movie Creator window so that you can use your new video content in a movie.

Making a Music Video with CineMagic

Roxio VideoWave Movie Creator actually has three different methods you can use to create a movie from your video content. In this section, we'll look at the first of these, CineMagic. This is a very simple, template-based movie creation method that is best suited to the most basic types of movies such as music videos. CineMagic doesn't offer a lot of sophisticated editing tools, but one thing that is does offer is absolute ease of use.

To begin creating your movie with CineMagic, click the CineMagic button on the main Roxio VideoWave Movie Creator window.

Naming Your Production

CineMagic suggests a default name for your movie using "CineMagic" followed by a number (which is incremented by one for each new video). You'll almost certainly want to enter a more descriptive name for your movie.

After you've specified the name that you want to use, click the Next button to continue.

Adding Your Video

Next you add the video clips that you want to include in your movie. As Figure 12-5 shows, you do so by dragging and dropping clips from the library into the box below the viewer window. The total length of your movie must be at least twice as long as whatever music you'll add, so that CineMagic has enough video to work with.

To the right of the box where you drop the videos are three controls. You can click the upward-pointing arrow to move a selected clip up in the play list. Clips that are higher in the play list will appear earlier in your video than ones that are lower in the play list. You can use the downward-pointing arrow to move clips down in the play list, and you can use the X to delete selected clips from the play list.

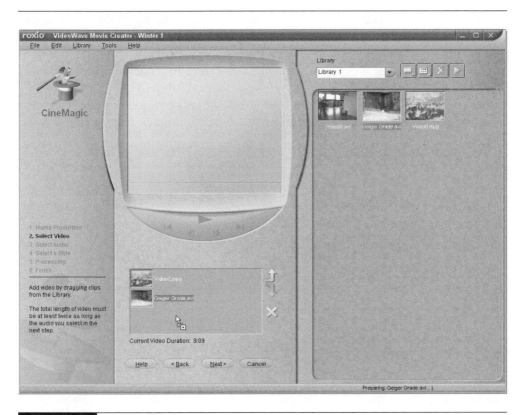

FIGURE 12-5   Select the video clips that you want to be a part of your music video.

Click the Next button once you have added all of the video clips that you want to include in your movie.

Adding Your Audio

You can use almost any Windows-compatible recorded audio file as the music track for your video. This could be a WAV file, an MP3, or even a track from an audio CD.

NOTE   *The one major type of audio file that you cannot use is a MIDI (Musical Instrument Digital Interface) file. MIDI files are not recorded audio; rather, they are instructions that tell the audio synthesizer in your PC how to create music. MIDI files can be played and recorded on your PC, but that process is a bit too complicated to discuss here.*

To add an audio track to your movie, you must first add that audio to the Roxio VideoWave Movie Creator library. Click the Add file(s) to Library button to display the Add files to Library dialog box, as shown here. If necessary, navigate to the folder that contains the files you want to add, select the files, and click Open. Don't forget that the audio track you add must be no longer than half the length of the video clips. For CineMagic to be most efficient, Roxio recommends that the video clips' total be close to three times the length of the audio track that you add.

Drag the audio track from the library into the area below the viewer window. This step may seem a little confusing at first, since Roxio VideoWave Movie Creator doesn't provide any sort of list box as the destination for your audio file. However, once you give it a try, you'll see that you can drop the file anywhere in the area directly below the viewer window.

Click the Next button after you have added your audio track.

Choosing a Style

CineMagic will now present you with a list of styles you can choose for your movie. These styles are editing styles rather than the menu styles you've seen elsewhere. For example, the Action 1 style is described as "a fast-cut style that's highly responsive to the music. Re-sequences material from the original video according to the music. Simple high-energy impact." You can see the description of each style by selecting the style in the list box below the viewer window.

> **TIP** *Remember that you can always click the Back button if you try out a style and decide it's not the one for you. Then choose a different style and click Next to preview it.*

Once you've found a style whose description sounds like it fits your video, click the Next button to continue. When you do, CineMagic will analyze all of the components of your video and generate a storyline based on the editing style that you have selected.

Completing Your Video

The video that CineMagic produces will likely be quite unlike anything you would create on your own. As I mentioned earlier, CineMagic creates a video production that is essentially a music video. In doing so, it uses sections that it cuts from the video clips you've added and matches them to the music. As you can probably guess, this automated video editing method can produce some rather unusual results. You will definitely want to preview your movie (as shown in Figure 12-6) before you burn a disc.

If you are satisfied with the results after you've previewed your movie, make certain that the Make Movie option is selected and then click the Finish button. You will then be presented with the following set of output options:

- **DV Camcorder**   Choose this option if you want to copy the video to a mini DV tape in your camcorder.

- **TV/VCR**   This option enables you to copy the video to an analog output so that you can record it on a videotape or watch it on a TV.

- **Internet**   If you want to put your music video onto a Web site, choose this option and you'll be able to select an appropriate format (depending on things such as download speed and video player software).

- **Prepare for DVD/VCD**   This option enables you to burn your video onto a recordable DVD or CD-R.

- **Video File**   This final choice produces a video file for viewing on a PC.

I won't go into the details of producing your final video here, since we'll be looking at this subject in more detail later in this chapter. If you do decide to

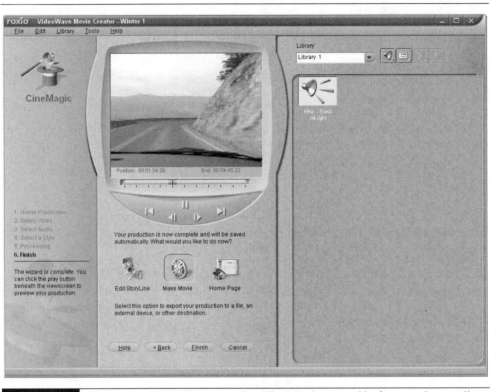

FIGURE 12-6 Preview your movie to see what CineMagic created before you burn a disc.

complete your music video at this point, simply follow the onscreen directions. Even if you stop at this point, you'll be able to come back to your music video later by using the *Storyline Editor* (which I discuss later in this chapter).

Using StoryBuilder

If you want a bit more control over your final movie than what is provided by CineMagic but still like a lot of handholding, you'll want to give the Roxio VideoWave Movie Creator *StoryBuilder* a try. This section of the program enables you to lay out your movie as you see fit, but it still provides a lot of help, so you won't have to do any more work than necessary. In most cases, you will probably find that StoryBuilder is a better choice for creating movies than is CineMagic.

To begin using StoryBuilder, click the StoryBuilder button on the Roxio VideoWave Movie Creator main window. Then follow along with the subsequent sections to produce your movie.

Setting Up Your Movie

As I mentioned, StoryBuilder provides a lot of assistance as you step through the process of creating your movie. When you first open the StoryBuilder, you'll be asked to perform a few basic setup tasks:

1. First, enter a name for your movie in the text box that appears on the first StoryBuilder screen. Remember to make this descriptive so that you'll be able to easily identify the movie project later.

2. Click the Next button to view the list of templates you can use. You may need to insert the Roxio VideoWave Movie Creator disc if you want to use a template that is currently grayed out.

3. After you have selected a template, click the Next button to view the available introduction screens. These are title screens that will appear at the beginning of your movie, and the selections are based upon the template that you selected.

4. To preview an introduction screen, click the green Play button just below the center of the viewer window. In this case, I'm previewing one of the intros from the Winter template. (Don't worry about the title on the intro screen for now—I'll show you how to change that next.)

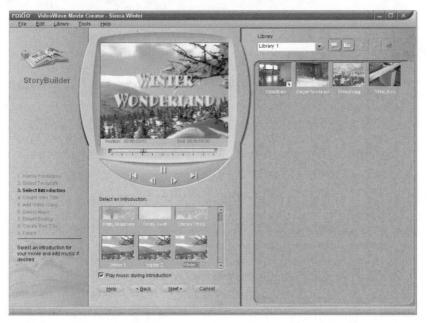

5. When you have selected your template and intro screen, click the Next button to continue.

Adding an Opening Title

In most cases, you will probably want to modify the opening title. The default titles that are suggested by StoryBuilder are surprisingly good compared to the simple placeholders most digital video editors offer—but even so, there's nothing like a title you personalize yourself.

To modify the opening title, follow these steps:

1. Enter your title in the text box that appears in the area below the viewer window, as shown here:

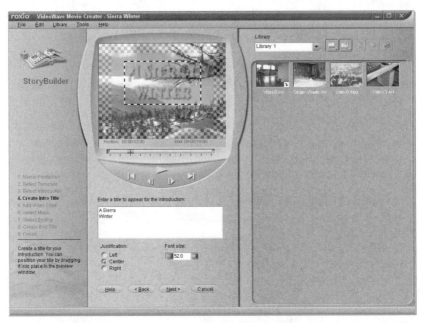

2. If you want to change the text alignment or the font size, use the options that appear below the text box to modify these properties.

3. If necessary, drag the text box in the viewer window to position it where you want it. Be sure to keep the text box within the TV safe zone (the area which is bounded by the checkered pattern)—otherwise someone viewing your movie on a TV may not be able to read the title, since it may be cut off.

4. Click the green Play button to preview your title. Notice that, as the title text box moves through the areas outside of the TV safe zone, the checkered pattern will change to yellow and black to indicate that there may be a problem reading the title while it is in that area.

5. When you are satisfied with your title, click the Next button to continue.

Adding Your Content

Next you have the opportunity to add your video and audio content to your movie. This, of course, is the fun part where your movie actually starts to take shape. In addition to being where you add content, this is where you can trim video clips if you want.

Let's take a look at how you add your content to your movie in StoryBuilder:

1. Drag-and-drop movie clips from the library into the list box below the viewer window. Notice that if you select a video file in which Roxio VideoWave Movie Creator has detected the scenes, the lower section of the library will show the individual scenes, as shown next. In such cases, you can add individual selected scenes rather than the entire video clip (if that's what you want, of course).

2. To preview a video clip, click the green Play button below the center of the viewer window.

3. If you want to trim the beginning or the ending of a clip, first select the clip you want to trim (in the list box) and then click the scissors button just to the right of the list box. This will display the controls shown in Figure 12-7.

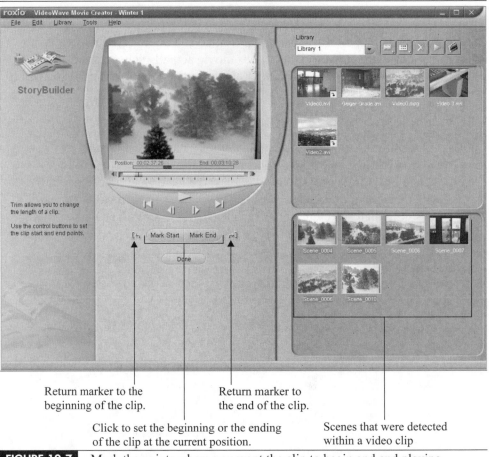

Return marker to the beginning of the clip.

Return marker to the end of the clip.

Click to set the beginning or the ending of the clip at the current position.

Scenes that were detected within a video clip

FIGURE 12-7 Mark the points where you want the clip to begin and end playing.

4. Use the playback controls below the viewer window to move to the point where you want the clip to begin playing, then click the Mark Start button to mark that point.

5. Move to the point where you want the clip to end and click the Mark End button to mark that point.

6. When you are finished trimming the clip, click the Done button.

7. Use the up and down arrows next to the right side of the list box if you want to rearrange the order of the clips.

8. Click the Next button when you are done adding and editing the video clips to move on to adding your background music.

9. Select a background music track from the list box below the viewer window. You can preview the selected music track by clicking the green Play button.

10. Click the Next button when you have finished selecting your background music track.

Adding an Ending

Next select an ending screen and create an ending title by following along with the onscreen prompts. The ending screen and ending title are created using the same steps you used earlier to add an intro and an opening title, so we'll skip over those steps at this point. I'm sure that you will have no problems adding these elements to your movie (but if you need a little more help, refer back to the "Setting Up Your Movie" and "Adding an Opening Title" sections earlier in this section).

Be sure to click the Next button when you have finished selecting an ending screen and adding a closing title.

Finishing Your Video

Your movie is now almost complete. When you reach this point, you'll want to click the green Play button to view your finished video. Roxio VideoWave Movie Creator automatically saves your completed movie file, so for now you can return to the main Roxio VideoWave Movie Creator window rather than burn the disc. As I mentioned earlier, we'll look at the disc production process later in this chapter.

Enhancing Your Movie with StoryLine Editor

The third method of assembling and editing your movie in Roxio VideoWave Movie Creator is to use the StoryLine Editor. This is by far the most flexible of the three methods, and it's the one that I feel you'll probably find most useful. In the following example, we'll use StoryLine Editor to open and modify the video we created in the last example (using the StoryBuilder).

Starting Your Production

The StoryLine Editor provides you with more direct access to Roxio VideoWave Movie Creator's editing tools than do either of the other two methods of creating a movie in the program. Therefore, StoryLine Editor does not provide the step-by-step wizard-type of interface those methods offer, and you can work on a video in StoryLine Editor in pretty much whatever order suits you. This is especially pertinent when you are editing an existing movie project, as we'll do in this example.

To begin working in the StoryLine Editor, click the StoryLine Editor button on the main Roxio VideoWave Movie Creator window. When you do, you'll see the dialog box shown here:

Choose the option that suits your needs. In this case, I've selected the option to open our existing production entitled "Winter 1" from the last example. Click the OK button after you've made your selection.

Adding and Editing Your Video

Figure 12-8 shows the StoryLine Editor window that you use to create and edit your movies. As you can see, this window places a lot of controls where you can easily access them (in this case, I've selected a video clip on the storyline so that the clip editing controls appear along the right side of the viewer window).

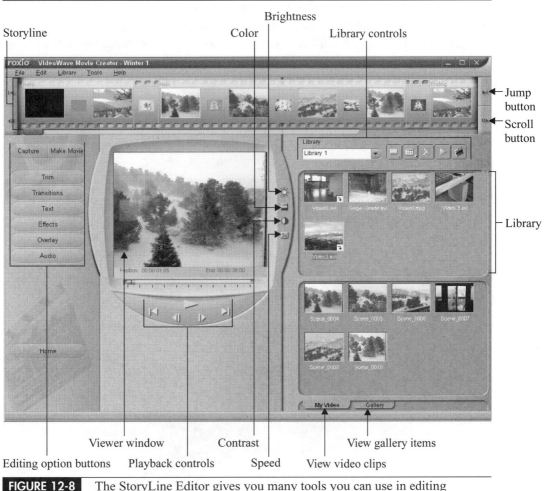

FIGURE 12-8   The StoryLine Editor gives you many tools you can use in editing
your movies.

Adding Clips

To add video clips to your movie, drag them from the library and drop them into
position on the storyline. When you drop the clip onto the storyline, any existing
clips automatically move to accommodate the new arrival. You can also drag clips
to different locations on the storyline; once again, the other clips will adjust their
position to make room for the clip you move.

If you're using imported video clips in your movie rather than ones you have
captured from your DV camcorder within Roxio VideoWave Movie Creator, you

may discover there is a slight problem with those video clips—there's no scene detection within those clips. (This can also be a problem with video clips you capture from an analog source.) It isn't really a problem if you want to use the entire clip within your movie, but it does make things awfully hard if you want to be able to pick and choose scenes from the clip. The solution to this problem is to ask Roxio VideoWave Movie Creator to attempt to detect scene changes within the clip. Then you'll be able to select individual scenes to add to your movie instead of being forced to add the whole clip and trim it after it has been added to the storyline.

Fortunately, Roxio VideoWave Movie Creator does a fairly good job of scene detection. To detect the scene changes in a clip, follow these steps:

1. Select the clip in the library. If Roxio VideoWave Movie Creator has already detected scene changes in the clip, you'll see a list of the scenes in the lower section of the library, as shown in Figure 12-8. If no list of scenes is shown, Roxio VideoWave Movie Creator has not detected any scenes.

2. Click the Scene detect button (the right-most button above the library) to display the Scene Detection dialog box shown here:

3. Click the Start Detection button to begin scene detection. When you do, Roxio VideoWave Movie Creator will play back your clip at high speed and attempt to detect the different scenes based on changes in the video content.

4. If necessary, drag the Detection Sensitivity slider to the left to reduce the number of scenes that are detected or to the right if too few scenes were detected. Then click the Start Detection button to begin scene detection again.

5. When you're satisfied with the scene detection results, click the OK button to close the dialog box.

> **TIP** *Even if the scene detection process doesn't produce quite the scene breaks you really want, you can still add the detected scenes and then trim them on the storyline to get right down to the exact frames you want to include in your movie.*

Trimming Clips

In Figure 12-7 you saw how to trim video clips in StoryBuilder. In the StoryLine Editor the process is similar (since you use the same controls as in StoryBuilder), but you have more flexible options. There is, however, one very important difference in the processes of trimming video clips in StoryBuilder and in StoryLine Editor—in StoryLine Editor, you trim clips after they have been added to the storyline.

This difference in methods is important because when you edit clips in the storyline it is much easier to see how your edits will affect the whole movie. In addition, since you can add the same clip to the storyline more than once, in StoryLine Editor you can easily make a single scene appear quite different each time it appears simply by trimming it differently. This capability provides you with far more precise control and flexibility in producing your finished movie.

As in StoryBuilder, trimming clips in StoryLine Editor consists of moving to the frame where you want the scene to begin playing and clicking the Mark Start button. Likewise, you move to the frame where you want the scene to end playing and click the Mark End button. Click the Done button when you are finished.

> **NOTE** *Trimming a clip does not affect the original copy of the clip that is stored on your hard drive. Trimming affects only how much of that clip is included when you finish your movie.*

Adding Transitions

Transitions are one of those movie elements that no movie actually needs, but they certainly can add a lot of visual interest when they are added to a movie. Transitions seem most effective when they are used to emphasize the differences between one scene and the next.

To add (or edit) a transition in StoryLine Editor, follow these steps:

1. Click the Transitions button to display the available transitions, as shown here:

2. To add a transition to the storyline, drag-and-drop the transition into the space between the two video clips that you want to come before and after the transition.

3. To view the transition in action, make certain that the transition is selected on the storyline and then click the green Play button (below the viewer window).

4. To change the duration of the selected transition, click the Edit button and then drag the duration slider right or left (as appropriate) to adjust the length.

5. To remove the selected transition, click the Delete button.

6. To replace a transition with a different transition, drag-and-drop a new transition onto the transition that you want to replace.

7. Click the Done button when you are finished.

Adding Special Effects

Roxio VideoWave Movie Creator also offers a number of special effects that you can apply to your movie. In Figure 12-9, for example, I've added the Search Lights special effect to one of the video clips in the movie.

Special effects are similar to transitions, but there are some important differences between the two:

- Special effects are applied to a single video clip rather than to the change between two clips.

- You can set a special effect to start or end at a different point than the video clip itself. That is, you can mark the point in a video clip where you want a special effect to begin, and you can mark a point where you want the special effect to end.

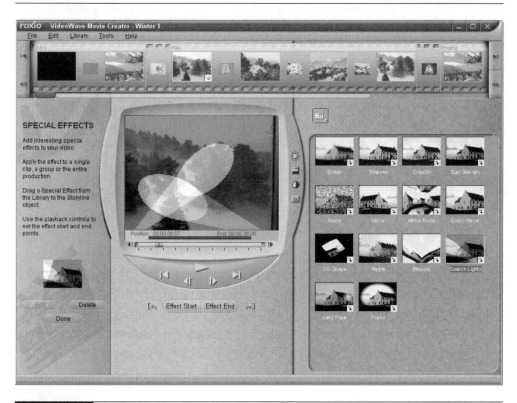

FIGURE 12-9   Special effects can be a lot of fun.

- You adjust the duration of a transition by using the duration slider, but this duration is generally limited to a few seconds. Special effects can run the full length of the video clip to which they are applied.

You add a special effect to a movie clip in much the same way as you add a transition to the storyline. Basically, you click the Effects button, drag-and-drop a transition onto a video clip, mark the beginning and ending points (if you want), and then click the Done button.

Adding Overlays

Overlays are yet another type of special effect you can add to your movies. In fact, they are essentially static special effects. That is, an overlay is simply a still image that is added on top of a video clip (such as a bunny you might add to a baby video).

Since overlays are so similar to special effects, I'll leave it up to you to play around with overlays on your own. My best advice regarding overlays is simply to use some restraint, since overuse of them can quickly become very annoying!

Adjusting the Video Properties

If you've ever shot a video in less than ideal conditions, you will probably appreciate how easy it is to adjust the properties of a video clip in the StoryLine Editor. Whenever you are playing a video clip in the viewer window of the StoryLine Editor, there are four controls along the right side of the viewer window that you can use to change the properties of the clip:

- **Brightness**   This control displays a slider that you can drag to make the selected clip brighter or darker, as need be. You could, for example, make a scene darker to give the impression that it was shot at night.

- **Color**   This control displays a slider that you can use to adjust the color saturation of the selected clip. You might want to increase the color saturation to make a scene shot on an overcast day appear a bit more alive.

- **Contrast**   When you click this control, you can use the slider to increase or decrease the contrast between light and dark areas of the scene. Decreasing the contrast slightly can bring out more of the detail in a scene you shot on a very bright day at the beach.

■ **Speed**   This final clip property control displays a pop-up menu that enables you to adjust the playback speed between one-eighth and four times normal speed. Speeding up the playback can be a good technique to simulate the passage of time.

> **TIP**
>
> *The Brightness, Color, and Contrast sliders all include a Reset button you can click to return that property of the movie clip to the neutral value where it started. If you don't feel the property adjustments are producing the result you want, clicking the Reset button takes you back to the point before you started tinkering. Remember, though, that each Reset button resets only the property that is adjusted by its own slider.*

Adding Titles

I doubt that very many people would argue with the fact that titles have been a part of movies far longer than many of the other movie elements we take for granted these days. Long before color and even sound were introduced, titles were an important part of the movies. Titles can serve a number of purposes in your movies, so it's good to know that the StoryLine Editor makes it so easy for you to add the titles that will put that finishing touch on the movie.

Adding titles to your movies in Roxio VideoWave Movie Creator will be a very easy process for you. If you have followed along with the earlier examples in this chapter, the steps should seem quite familiar. Here is how you can add titles to your movies in the StoryLine Editor:

1. Click the Text button to the left of the viewer window to view the text tools.

2. Select a text style in the library area. In this case, select the Motion tab, then select Fade In in the upper pane, and then select Fade In And Out in the lower pane. The lower pane appears when your selection in the upper pane has more than one variation.

3. Drag your selection onto the video clip where you want the title to appear.

4. Click the Edit button to open the text editing tools, as shown here:

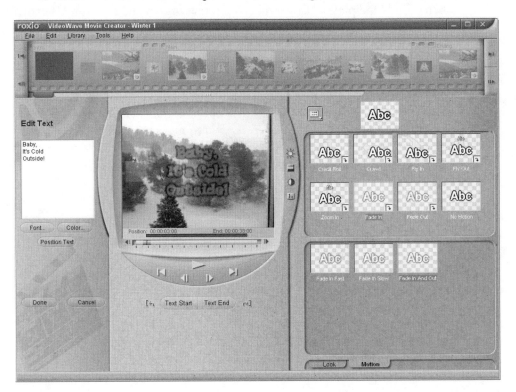

5. Enter your title in the Edit Text box to the left of the viewer window.

6. To choose a different typeface, click the Font button and make your selection.

7. To choose a different color for the text, click the Color button and select the color you want.

8. If you want to move the title, click the Position Text button and drag the text box to the desired location. Notice that this button acts as a toggle, and it also displays the checked pattern to indicate the TV save zone.

9. If you want the title to be displayed some time after the beginning of the video clip, move the playback position to the point where you want the title to begin and click the Text Start button.

10. If you want the title to end at some point before the end of the video clip, move the playback position to the point where you want the title to end and click the Text End button.

11. Click the green Play button to preview your title.

12. Click the Done button when you have finished editing your title.

13. Click the next Done button when you have finished creating new titles.

You can add titles to any of the video clips in your movies, of course. Keep in mind that the old-time moviemakers used titles quite effectively to help tell the story when they didn't have recorded soundtracks. This same technique might be very effectively used in some of your movies, too.

Adding Audio Content

By now you should be pretty familiar with Roxio VideoWave Movie Creator and the StoryLine Editor. You've learned how to add various elements to your movies, and you've seen that the methods you use in each case are consistent throughout. It should, therefore, come as no surprise to learn that adding audio content follows this same path.

When you click the Audio button, you'll be presented with a view similar to the view shown in Figure 12-10. I've clicked on the Loops tab of the library to give you a look at some of the types of musical selections you can add to your movies.

NOTE | *In Roxio VideoWave Movie Creator, you have the choice of both clips and loops if you want to add royalty-free background music to your movies. The only important difference between clips and loops is simply that loops automatically repeat as long as necessary until the video clip finishes playing.*

You can, of course, use tracks you've ripped from audio CDs as well as most other Windows-compatible audio files as background music for your movies. Simply select the My Audio tab to choose one of those files.

There are three methods for adding audio to your movie. Choose the option that best suits your needs from the following choices:

■ To add an audio track to an individual video clip, drag-and-drop the audio clip onto the scene's thumbnail on the storyline.

■ To add an audio track to a series of contiguous video clips, click the first video clip in the series, hold down SHIFT, click the final video clip in the series, and then drag-and-drop the audio clip onto the selected video clips.

- To add an audio track to the entire movie, right-click the storyline and choose Select Production from the pop-up menu. Then drag-and-drop the audio clip onto the storyline.

If you want to adjust the length of an audio clip that you've added to the storyline, click the Edit button while the clip is selected. You will then be able to mark the starting and ending points for the audio track (in the same way you marked the starting and ending points for titles, for example).

While you are editing the audio track, you can also use the volume control sliders to adjust the audio levels. In addition, you can choose to make the audio track loop, fade in, or fade out. Be sure to click Done when you are finished making your adjustments.

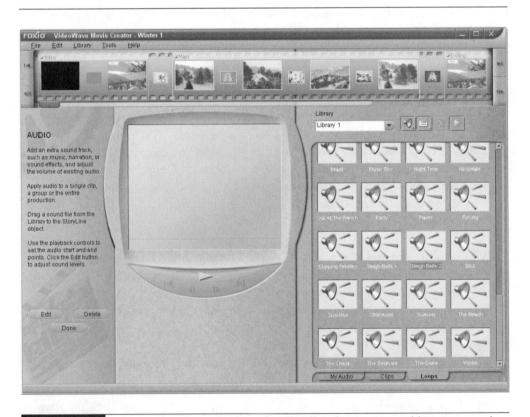

FIGURE 12-10   There are quite a number of audio clips that you can add to your movie.

Preparing Your Movie for Output

When you have finished editing your movie in the StoryLine Editor, you are ready to begin the process of choosing an output format and rendering your movie. Doing this produces the file that can later be burned to a disc.

To prepare your movie for final output, follow these steps:

1. Click the Make Movie button to display the output options, as shown here:

2. Select the type of output you want to produce. In this case, select the Prepare for DVD/VCD option.

3. Click the Next button.

4. Choose the Purpose and Quality options that suit your needs. In this case, select DVD playback and Normal quality.

5. Click the Next button to continue.

6. Enter a file name for the output file. This can be the same as your project name, since the output file will have a different extension than the project file (which means that it won't overwrite the project file).

7. Click the Next button to continue.

8. Click the Render file button to begin creating the output file. Depending on the performance of your computer and the size of the movie, the rendering process may take considerable time to complete. I suggest not using your PC for anything else while the rendering is taking place, since other tasks will slow down the rendering.

9. When the rendering is complete, you can click the Preview button to view your finished movie.

10. Click the Prepare Disc button to move on to the VideoCD/DVD Maker. Alternatively, you can return to the main Roxio VideoWave Movie Creator window and burn your disc at a later time.

> **TIP** *If you want to include more than one video on your disc, prepare all of the videos that you want to include before you move on to the disc burning process.*

Burning Your Disc with VideoCD/DVD Maker

We're finally down to the last major task in creating your movie with Roxio VideoWave Movie Creator. We will now have a look at how you can burn your disc so that you'll have something to show for all of your effort.

> **NOTE** *The VideoCD/DVD Maker option on the main Roxio VideoWave Movie Creator window is strictly for creating discs. If you want to create a different type of output—copying your movie to a videotape, for example— you need to use the Make Movie button in the StoryLine Editor to choose one of the non-disc movie formats.*

Opening Your Project

To begin the process of creating your disc, click the VideoCD/DVD Maker button on the main Roxio VideoWave Movie Creator window. If you chose to continue on from the Make Movie wizard in the previous section, you'll already have the VideoCD/ DVD Maker section of the program open and won't need to click the button.

At this point you may be slightly confused by your options. As shown in Figure 12-11, the video that you just prepared is not offered as the project name. Don't let this throw you. If you look closely at the library, you'll see that the movie project that you just created is indeed sitting there in the library. You'll be able to add it to your disc shortly. For now, enter a name for your disc and then click Next.

Choosing a Disc Format

Now you have the opportunity to choose the type of disc you want to create. The DVD and VCD options should be pretty easy to understand, but the DVD on CD option may seem a little unusual. Actually, this is just another name for S-VCD, which we have discussed several times earlier in this book.

TIP	*If you want to create a disc for playback in a set-top DVD player, DVD is your best choice. A VCD is not quite as compatible with set-top boxes, and S-VCD (DVD on CD) is the least compatible.*

Select your disc format and then click Next to continue.

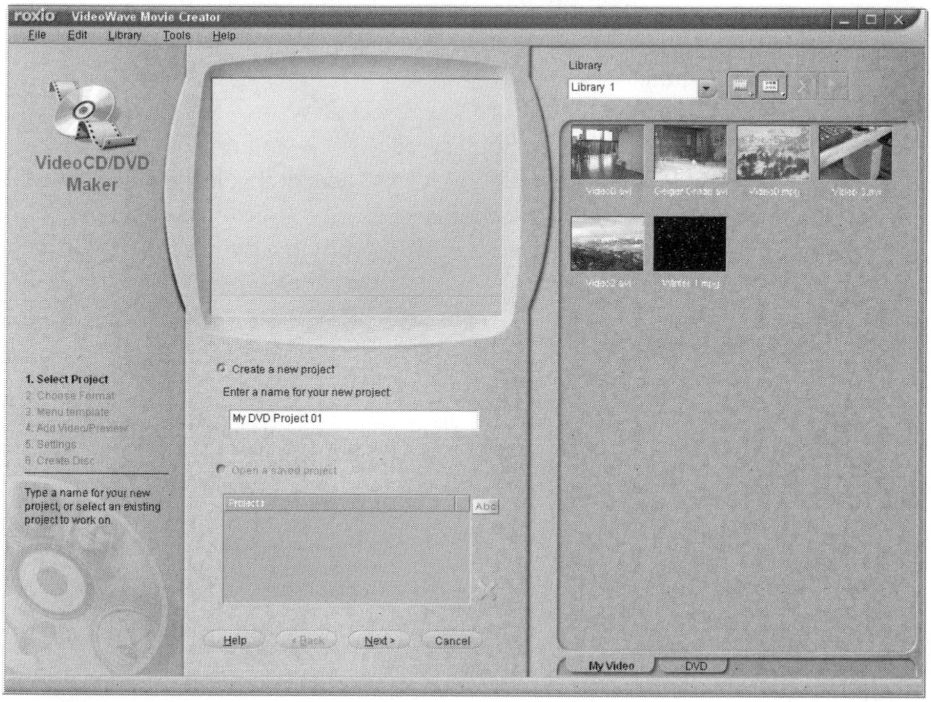

FIGURE 12-11 Look in the library to find your new movie project.

Selecting a Menu Template

Select a menu template for your disc from the list below the viewer window. Roxio VideoWave Movie Creator offers a number of templates, but you don't have the option of creating your own menus—you'll have to be satisfied with one of the choices shown in the list.

If you want your DVD to play music whenever the menu is displayed, select the "Include audio with menu screens" checkbox. Remember that menus automatically loop, so the music will play constantly until the person viewing your movie makes a selection.

Click the Next button after you've selected your menu template.

Adding New Video and Finalizing Your Settings

At the Add Video/Preview screen, drag-and-drop your prepared movie files into the box below the viewer window, as shown in Figure 12-12. Be sure to drag-and-drop only files that have an MPG file extension—other types of files have not been rendered in the proper format for inclusion on a DVD yet (see "Preparing Your Movie for Output" earlier in this chapter).

Use the controls along the right side of the video list box to select a thumbnail, rename an item, or change the order of the items. When you have finished, click the Next button to continue.

Since you likely have only one recordable DVD drive on your system, the only option you'll be able to select on the setting screen will be to preview the final results before burning the disc. It's generally a good idea to accept this last chance to make certain everything is okay before you actually commit to burning a disc. (You may see other options if your system is configured differently or if you selected different output options.)

Burning Your Disc

You're almost finished. At this point, you can click the Create Disc button to instruct Roxio VideoWave Movie Creator to begin creating the disc image—essentially a copy of the final disc that is stored on your hard drive before the disc is burned. After you click the Create Disc button, you'll see progress indicators, as shown in

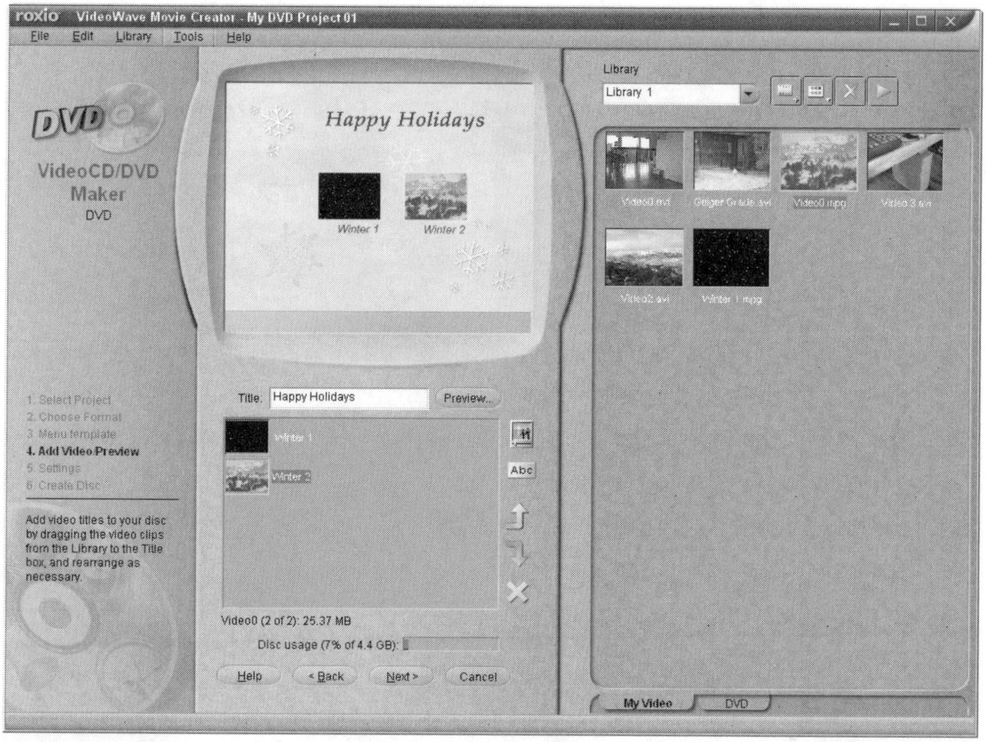

FIGURE 12-12   Use the box below the viewer window to organize your movie.

Figure 12-13. Wait until the disc image has been completely created before you do anything else.

I strongly suggest that you take the opportunity to preview your disc once the disc image has been created. Use the onscreen remote control to test your disc menus. When you are satisfied, make certain that you have a blank, recordable disc of the proper type for your project in the drive and then click the Burn Disc button. Since Roxio VideoWave Movie Creator has already created a disc image on your hard drive, actually burning the disc will be a pretty simple and straightforward process.

In this chapter, I showed you how to create a complete movie using Roxio VideoWave Movie Creator. As you have learned, this program offers three different methods of preparing your disc content, and the one you choose really depends on

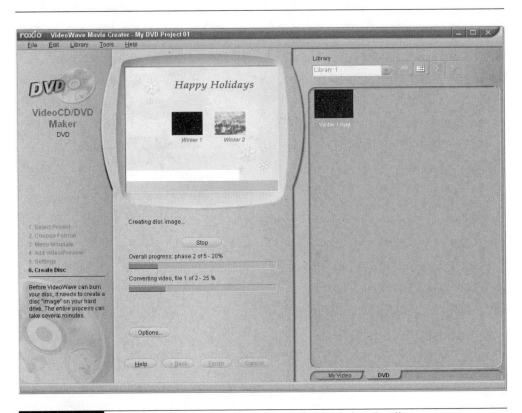

FIGURE 12-13 My movie is being prepared for being added to the disc.

your needs. Each offers a different approach to creating a disc, each offers varying amounts of handholding, and each offers a different mix of options for customizing your videos. You'll probably find yourself playing around with each of the three methods and then settling on one of them for most of your movies.

In the next chapter, we'll move on to the next level of digital video editors with coverage of Sonic DVDit! As you will see, the mainstream digital video editors trade some of the handholding you've seen in these first three editors for more flexibility and power. I think you'll find that the capabilities that these editors provide are well worth the small amount of extra effort when you want to produce the best video possible.

CHAPTER 13

Creating a Movie in Sonic DVDit!

S *onic DVDit!* is quite different from most of the other programs we've been discussing throughout this book. The primary difference lies in the focus of Sonic DVDit!. Unlike the other programs you've seen so far, Sonic DVDit! is primarily a DVD publishing application rather than a digital video editing one. That is, Sonic DVDit! enables you to take a video you have captured and edited in another program and create the fancy menus and navigation options for it that you might find on a professionally produced DVD.

It is important to understand that Sonic DVDit! is not really intended as a standalone, complete movie production system. It was designed to assemble the input from a number of sources and help you complete the production as DVD output. Since DVDit! is not intended to be a video capture application, you must use video files that you have captured with your other input applications. These might be programs such as the video editors contained in Sonic's MyDVD Video Suite, ArcSoft ShowBiz, Adobe Premiere, Microsoft Windows Movie Maker, Avid DV Express, or even a simple video capture program that came with your video capture hardware or camcorder. In addition, Sonic DVDit! does not include the tools for editing your video clips after they have been captured. Here, too, the intent is that you will use one of these other programs to prepare the video, which you will then import into Sonic DVDit! and add your menus.

NOTE *During discussions with the people at Sonic, I was told that future versions of Sonic DVDit! will add features such as video capture and some video editing capabilities. You may want to check the Sonic Web site (www.estore.sonic.com) to learn about possible version upgrades if these features are important to you.*

Starting a New Project

When you open Sonic DVDit!, you must answer a series of questions before you can begin using the program. As this shows, you first decide if you want to start a new project, work on an existing project, or exit from the program:

Next you choose the TV standard and video format that you want to use, as shown here:

For American TV sets, always choose NTSC (PAL is the European standard). To create a DVD for DVD players, choose the MPEG2 (DVD-compliant) option. Click the Next button after you have made your selections.

Finally, choose a screen size, as shown here:

In most cases you will want to choose 4:3 (Typical TV)—this is the format used by most camcorders. If you have wide-screen video, you can choose the 16:9 option. Click the Finish button to confirm your choices and begin using Sonic DVDit!.

Understanding the DVDit! Screen

The Sonic DVDit! window is quite simple and easy to understand. Figure 13-1 shows the view that you will see when you start a new project and identifies the important elements in the window that you will use.

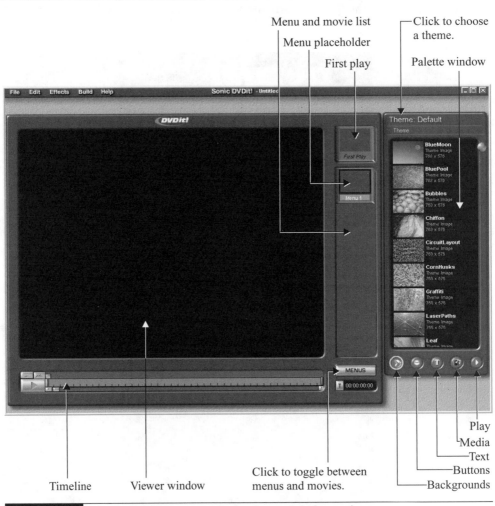

FIGURE 13-4   Sonic DVDit! presents a simple, straightforward appearance.

Let's take a closer look at the elements of the Sonic DVDit! window so that you can gain a better understanding of their purpose:

■ The timeline and the viewer window are comparable to their counterparts in the other programs we've been discussing, so they really don't require any further description.

■ The menu and movie list to the right of the viewer window is essentially a storyboard. Each item in this list appears in the movie in the order that it appears in the list (starting at the top).

■ The *first play* is the term for the first thing someone will see when the DVD is inserted into the DVD player. This can be a menu or a movie. If you delete the first play object, the DVD will not start playing automatically, and the person viewing the movie will need to use their remote control to begin the playback.

■ The button that is labeled "Menus" in Figure 13-1 toggles between adding menu objects and movie objects to the menu and movie list. This may be a little a confusing to deal with when you first start using Sonic DVDit!, because you will need to remember to click the button labeled "Menus" when you want to add a movie clip, and the same button (which after toggling is labeled "Movies") when you want to return to adding menus to the list.

■ The palette window displays selections of menu backgrounds, menu buttons, text objects, or movie clips, depending on which of the buttons at the bottom of the window is selected. (The Play button below the palette window previews the currently selected object.) You drag-and-drop objects from the palette window into either the viewer window or the menu and movie list, as appropriate.

■ Sonic DVDit! organizes movie elements into themes. You can choose a theme for your movie by clicking the Theme selector at the top of the palette window and then making a selection from the menu. You can also create new themes containing elements that you add (which might be a good idea to do if you are creating a series of movies that should share a common appearance).

Importing Your Video

In order to create your movie, you need to begin by importing your video files. Sonic DVDit! can use files in a number of different formats (although it will sometimes complain that a file that was produced by another manufacturer's software is not in the proper format, since some MPEG files imported from other applications are not DVD legal).

To import your video files, follow these steps:

1. Capture your video using your favorite video capture software and save it on your hard disk.

2. If you wish to edit the video clips or add special effects, such as transitions between scenes, do so in your digital video editing program of choice. Your goal is to produce video clips that have all of the video and audio effects you want in them completed before you import them into Sonic DVDit!.

3. Click the Media button below the palette window in Sonic DVDit! to open the media palette.

4. Right-click the media palette and then click the single-item Add Files To Theme menu that pops up to display the dialog box shown here:

5. Select the files you wish to add and then click the Open button to add those files to the media palette. If you see an error message similar to this one, you will need to correct the problem using the application that created the video file before it was imported into DVDit!.

TIP *You can also drag-and-drop files from a Windows Explorer window onto the media palette if you find that method more convenient.*

Previewing Your Video Clips

You will likely find that you want to preview the video clips that you have imported into Sonic DVDit!. But if you try to do so, you may discover that there doesn't seem to be a method of playing those clips without first adding them to the menu and movie list. Fortunately, there is a method that works if you know how to find it.

As shown in Figure 13-2, to preview a movie clip that is contained in the media palette, you right-click the clip and then choose Play from the pop-up menu. In this case I am previewing a clip named Video 7.avi, and I am about to preview a different clip named Video 3.avi. If you are previewing one clip and want to preview a different one, you do not need to stop the playback of the first clip before you view the second one—playing the second clip automatically stops playback of the first clip.

Selecting a First Play

You have probably never given the subject any thought, but were you aware that DVD players have to be told what to do when a disc is inserted into the drive?

FIGURE 13-5   Right-click a clip in the media palette to display a pop-up menu.

If you have ever played a Hollywood movie in a DVD player, you have probably noticed that when you insert the disc you likely will see an FBI warning screen displayed for a few seconds and then either the movie starts to play or a menu appears. That first action is called a *first play*. In many ways, a first play is similar to the autorun command that tells a Windows-based PC what action to take when a disc is inserted into a CD-ROM drive.

You can add a menu, a video clip, or a still image as the first play. To add a menu, you drag a background from the background palette onto the first play placeholder at the top of the menu and movie list (we'll discuss more about creating menus shortly). To add a video clip or a still image, you drag the video clip or image file onto the first play placeholder.

> **TIP** *Adding a video clip or a still image to the first play actually creates a movie placeholder in the menu and movie list, but you won't see the placeholder until you click on the Menus button and choose Movies from the pop-up menu.*

You can add additional video clips to your movie in two different ways. In either method, you'll need to make certain that the button below the menu and movie list is showing the correct label:

- If the button shows Movies, you can drag-and-drop a movie clip onto the menu and movie list. Doing this creates a movie entry in the menu and movie list, but does not link that entry with a menu button.

- If the button shows Menus, you can drag-and-drop a movie clip onto a menu background or onto a button in the viewer window. Doing this creates a movie entry in the menu and movie list and also links the movie to a menu button.

Whichever method you use to add new video clips to your movie, Sonic DVDit! automatically adds a new, blank movie placeholder at the bottom of the list of movies.

Adding Menus

Menus are the navigational controls for your movies. Menus can be linked to video clips, to additional menus, and to chapter points within video clips. By creating links to chapter points you make it possible for viewers to jump to specific points within a video clip—not just to the beginning of the clip.

The menus you create in Sonic DVDit! are far more flexible than the menus you can create in any of the template-based editors we've looked at earlier. In Sonic DVDit! you have complete control over the layouts of your menus, and you can add up to 36 buttons on a single menu if you like (but I don't suggest that you go quite so far overboard, since too many buttons on a single menu can make navigation more rather than less difficult).

To create a menu, follow these steps:

1. Make certain that the button below the menu and movie list says Menus. If it does not, click the button and choose Menus from the pop-up menu.

2. Click the Backgrounds button at the lower left of the palette window to display the backgrounds palette.

3. Optionally, click the theme selector and choose a different theme to open from the menu.

4. Drag-and-drop the background image that you want to use for your menu onto the top-most empty menu placeholder in the menu and movie list. When you do, Sonic DVDit! will add an additional menu placeholder at the bottom of the list.

5. Be sure to frequently select the File/Save command from the Sonic DVDit! menu as you are working on your movie to save any changes you have made.

> **TIP** *When creating a menu, a "Photos To Go" logo will appear outside the TV safe zone, so it will not show up when your movie is played on a TV. If you wish to purchase menu backgrounds from Photos To Go without the logo, visit their Web site by using the Photos To Go link in DVDit!'s Help menu.*

At this point you have two options for creating the buttons for your menu:

- As I mentioned earlier, you can drag-and-drop movie clips from the media palette onto the menu background in the viewer window to create a menu button which links to the first frame of the selected clip. The button will use the first frame of the clip as a thumbnail.

- You can open the button palette and then drag-and-drop a button onto the menu background. This new button will initially not be linked to anything, so you'll have to add your links manually. Fortunately, adding the links manually is quite easy.

Figure 13-3 shows an example of a menu that has four buttons. Three of them were created by dragging-and-dropping video clips onto the menu. The fourth (the lower-left button) was created by dragging-and-dropping a menu button onto the menu. This fourth button will need to have a link added manually. Notice that you are not locked into a specific menu layout in the way you are in the template-based editors.

Editing Your Menus

You can make a number of changes to the menus that you create in Sonic DVDit!. Because your menus are not based on a pre-built template, you have great flexibility in creating just the menu appearance you want. In fact, the only real limitation on your menus is that they cannot contain more than 36 buttons on a single menu. I doubt that you will consider this to be much of a limitation!

Because Sonic DVDit! gives you so much flexibility, it's hard to imagine everything you might do with your menus. Let's take a look at some of the things that you are likely to want to do with them.

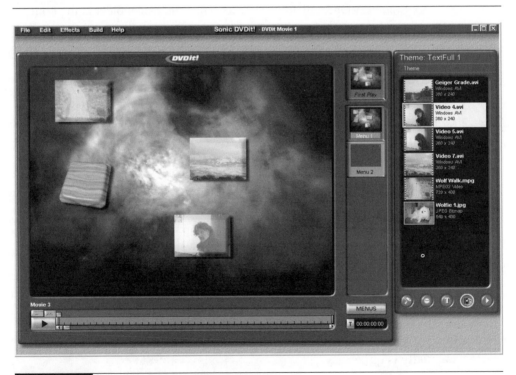

FIGURE 13-6   Here is a menu with four buttons.

Adding Chapter Points

As I mentioned earlier, chapter points are used for navigation within your movies. They provide a means of quickly jumping to a point other than the first frame of a video clip. Chapter points are a Sonic DVDit! feature that you will definitely want to take advantage of in many circumstances.

Each movie clip automatically has a first chapter, which starts at the first frame of the clip. You cannot move this chapter point. Any additional chapter points you add will be sequentially numbered starting with Chapter 2. Each movie clip's chapter points are independent of the chapter points in any of the other movie clips.

To add a chapter point to a movie in Sonic DVDit!, you double-click the timeline (while the movie clip is visible in the viewer window) at the point where you want to add a chapter point. Although this sounds simple enough, Figure 13-4 demonstrates why adding chapter points can sometimes be a little tricky.

Chapter point

FIGURE 13-7 This chapter point should be moved to a different frame to reduce the distortion.

You can move a chapter point to a better location in one of the following ways:

- Hold down CTRL as you drag the chapter point to a new location on the timeline.

- Click the chapter point to select it and then press the right or left arrow key to move the chapter point to the adjacent complete frame.

- Click the chapter point to select it, hold down SHIFT, and then use the right or left arrow keys to move the chapter point in one-second increments.

> **TIP** *It might be easier to add chapter points at specific locations by moving the timeline scrubber to the correct frame and then pausing the playback. Then right-click the timeline and select Insert Chapter from the pop-up menu to add the chapter point at the currently visible frame.*

Adding Manual Links

When we added buttons to the menu earlier, three of those buttons were linked to video clips, but the fourth button wasn't linked to anything. In other words, this fourth button is a non-functional button, since it won't do anything when it is clicked. We will need to add a link to the button if we want it to serve a useful function.

Incidentally, you can view the links that are attached to the menu buttons by right-clicking any of the buttons and choosing Show Button Links from the pop-up menu. In Figure 13-5 you can see that button 2 (in the lower left) does not have any links attached, while each of the other buttons is linked to a movie.

Adding a link to a button is a simple drag-and-drop process. You simply drag the object you want and drop it on the button to create the link. That object can be a movie clip, a chapter point, or another menu, depending on the type of link that you wish to create.

Adjusting Menu Colors

Another area you may want to modify is the colors that are used for the various parts of the menus. For example, you might decide that the background is too dark, that the color is too intense, or that the tint is just wrong for your movie. In such cases, you can easily adjust the colors using the Color Adjustment dialog box, as

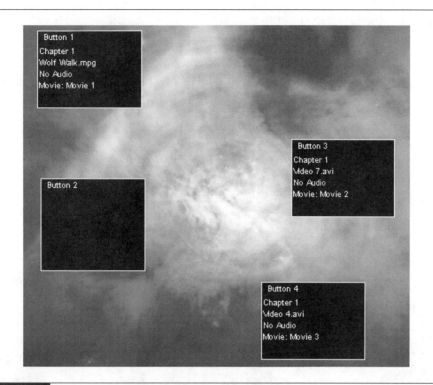

Button 1
Chapter 1
Wolf Walk.mpg
No Audio
Movie: Movie 1

Button 3
Chapter 1
Video 7.avi
No Audio
Movie: Movie 2

Button 2

Button 4
Chapter 1
Video 4.avi
No Audio
Movie: Movie 3

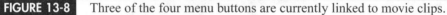

FIGURE 13-8   Three of the four menu buttons are currently linked to movie clips.

shown in Figure 13-6 (the dialog box is sitting near the right side of the Sonic DVDit! window in the area where the palette window would normally appear).

To use the Color Adjustment dialog box to adjust the colors, follow these steps:

1. Select Effects/Adjust Color from the Sonic DVDit! menu.

2. Select the object you wish to adjust from the drop-down Apply To list box. If you don't make a selection, your adjustments will be applied to the entire menu, including both the background image and the buttons.

3. Drag the hue slider left or right to adjust the color balance. You'll need to experiment with this control to get the colors you want.

4. Drag the saturate slider to the right to add more color or to the left to reduce the amount of color.

5. Drag the brighten slider to the right to make the object lighter or to the left to make it darker.

6. If you want Sonic DVDit! to use these same settings the next time you open the Color Adjustment dialog box, select the Save Settings checkbox.

7. Click OK when you are satisfied with the results to apply them and to close the Color Adjustment dialog box.

TIP *You can adjust a single object such as a button by selecting it before you open the Color Adjustment dialog box. This will enable you to correct the color balance of one menu button without changing any of the other objects on the menu. Be sure to select the "Selected menu items" option in the Apply To list box to limit the changes to the selected items.*

FIGURE 13-9 Use the Color Adjustment dialog box to tweak the colors to suit your movie.

Adjusting the Drop Shadow Effect

Objects that you add to Sonic DVDit! menus use *drop shadows* to give them a 3-D effect. You can use the Drop Shadow dialog box, as shown in Figure 13-7, to adjust these effects.

To adjust the drop shadow effects using the Drop Shadow dialog box, follow these steps:

1. Select the Effects/Drop Shadow command from the Sonic DVDit! menu to display the Drop Shadow dialog box.

2. Select the items you want the adjustments to apply to from the drop-down Apply To list box.

3. Drag the distance slider to the right to move the shadow further away from the menu objects or to the left to bring the shadows closer to the objects.

4. Drag the blur slider to the right to blur the edges of the shadows or to the left to make the edges sharper.

5. Drag the opacity slider to the right to make the shadows more opaque or to the left to make them more transparent.

6. Use the upper color slider to adjust the color balance of the shadows.

7. Use the lower color slider to control the brightness of the colors in the shadows.

8. Drag the white spot lamp symbol around the light source indicator to change the angle of the virtual light source (and therefore the angle of the shadows). When you do so, note the light source angles listed in the light source indicator. The zero point is on a horizontal line to the right of the indicator. The numbers range from 0 to 180 degrees above the horizon, and from 0 to -180 degrees below the horizon.

9. If you want Sonic DVDit! to use these same settings the next time you open the Drop Shadow dialog box, select the Save Settings checkbox.

10. Click OK to apply your changes and to close the dialog box.

> **TIP**   *Since you can change the drop shadow effects for various menu objects individually, it would be relatively easy to create a menu where each of the buttons had a drop shadow that radiated outwards from the center of the screen.*

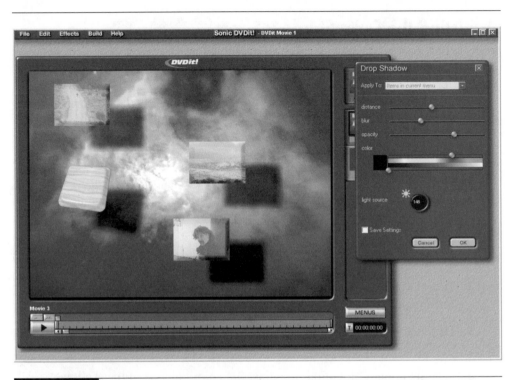

FIGURE 13-10 Drop shadows make it appear as though the objects were floating above the menu background.

Adding Titles

You will probably want to add some titles to your movies, so we will have a look at that topic next. Before we do, however, I need to remind you that we're talking about menu titles here, not titles that appear while your movie is playing. Sonic DVDit! does not provide any facility for adding titles to your movie clips, as this is usually provided by the video editing package that you used to import and edit your sources.

Sonic DVDit! is very flexible about allowing you to add titles just about anywhere you want in your menus. Even so, you should always remember to place any titles within the TV safe zone to ensure that they will be readable on any TV set. (Sonic DVDit! does not provide visual indicators to show you exactly where the TV safe zone is, so it's probably best to avoid the outside 10 percent or so of the viewer window when you are placing your titles on a menu.)

To add titles to your menus in Sonic DVDit!, follow these steps:

1. Click the Text button at the bottom of the palette window to open the text palette.

2. Click the typeface that you want to use for the title. Remember that you can use the scroll button along the right side of the text palette to locate additional fonts (you may have a set of fonts installed on your system that are different from those installed on my computer).

3. Drag the selected font into the viewer window to create a title that contains the word "text" (actually a title placeholder).

4. Click the title to select it, as shown here—note that I moved the menu buttons from their earlier positions to make more room for the title:

5. Enter your title text to replace the title placeholder. Don't worry if the text seems to disappear as you type; it will reappear completely when you have completed entering your title and click outside the text box.

6. To adjust the text properties, make certain that the text box is selected (if the text box does not have an outline with red handles, click it to select it).

7. Select Effects/Text Properties from the Sonic DVDit! menu to display the Text Properties dialog box, as shown here:

8. To choose a different typeface, make a selection from the drop-down Face list box.

9. You can ignore the Script drop-down list box since it is used to choose a different language set.

10. Use the text box or the drop-down list box in the Size section if you want to change the size of the text. Remember to keep the titles fairly large so that they will be easily readable on a TV.

11. Use the checkboxes in the Effects section if you want to make the text bold, italic, or underlined.

12. Drag the upper color slider to change the color of the text.

13. Drag the lower color slider to change the balance between black and white in the text.

14. Select the Save Settings checkbox if you want Sonic DVDit! to use these same settings the next time you open the Text Properties dialog box.

15. When you are satisfied with your selections, click the OK button to confirm your changes and to close the dialog box.

16. If necessary, drag the text box to position it properly. Some text property changes can result in a title that extends off the side of the screen.

17. Remember to select File/Save to save your work.

> **TIP** *If your changes have made the title too wide to fit on the screen, you can make your title into a multiline title by clicking in the title just before a word that you want to appear on a new line and then pressing ENTER.*

In addition to adding menu titles, you can add text to the menu buttons by dragging-and-dropping the text onto a button. This may be especially important if it is somewhat difficult to determine the content of each movie clip from the thumbnails that are displayed on the face of the buttons.

> **NOTE** *You can also change the thumbnails that appear on the buttons by selecting the button you want to change, moving the timeline scrubber until the frame you want to use is displayed, and then right-clicking the timeline. Choose Set Button Thumb from the pop-up menu to change the thumbnail.*

Adding Soundtracks

You can add a soundtrack to a menu or to a movie clip in Sonic DVDit!—although in most cases you will probably want to add a soundtrack to your movie clips using whatever editing program you used to create the movie in the first place. In either case, the procedures that you will use are very similar.

Preparing Audio Files

Before you can add a soundtrack to your movies in Sonic DVDit!, you will have to prepare the audio file by adding it to the media palette in the same way that you earlier added movie clips to the media palette. Sonic DVDit! is somewhat picky about the types of audio files that you can use for soundtracks. Your options are the following:

- Wave files with a sample rate of 48 kHz; a sample size of 8, 16, or 20 bits; a single audio stream, stereo sound, and a file extension of wav.

- MPEG-1 Layer II files with a sample rate of 44.1 kHz or 48 kHz; a sample size of 8, 16, or 20 bits; two-channel stereo sound (no rear channel); and a file extension of abs or mpa.

- Dolby Digital files with a sample rate of 48 kHz; a sample size of 8, 16, or 20 bits; two-channel stereo sound (no rear channel); a bit rate of 96 kHz to 448 kHz; and a file extension of ac3.

If you want to use another type of audio file (such as a track from an audio CD), you will have to convert the file to one of the preceding formats using another application before importing it into Sonic DVDit!. The digital video editor you used to capture and edit your movie files may offer this capability.

TIP	*You can also use the Windows Sound Recorder accessory to perform certain conversions for use with Sonic DVDit!. For example, you can convert a monaural wave file into a stereo one. Use the File/Properties command in Sound Recorder to convert audio files from one format into another one.*

Once you have audio files in the proper format for use in Sonic DVDit!, you can add them to the media palette. To do so, click the Media button at the bottom of the palette window to display the media palette. Then drag-and-drop your audio files from Windows Explorer into the media palette. You can also right-click the media palette and choose Add Files To Theme from the pop-up menu if you prefer that method. Either way, add the files you want to use with your menus or movie clips before continuing.

Adding Audio to Menus or Movie Clips

There are three methods for adding audio clips from the media palette to a menu or to a movie clip. You can use the option that best suits your needs from the following list:

- To add an audio track to a menu, make certain that the button below the menu and movie list shows Menus. Then drag-and-drop the audio clip from the media palette onto the menu placeholder. The audio track will then loop continuously whenever the menu is displayed.

- To add an audio track to a movie clip, make certain that the button below the menu and movie list shows Movies. Then drag-and-drop the audio clip from the media palette onto the movie placeholder.

■   Alternatively, you can drag-and-drop an audio clip from the media palette onto the menu button that is linked to the movie. You might want to use this method if you want the audio track to play only when a specific button is clicked (to add background narration, perhaps).

Sonic DVDit! is primarily an authoring tool, so any audio you add to your menus or to your movie clips will simply play when the menu or the movie clip is played. If you want to adjust the volume, for example, you'll need to do so before you import the audio file into the media palette.

Previewing Your Movie

When you have finished building your movie, you'll want to preview it, as shown in Figure 13-8. This will enable you to verify that everything works the way you expect and that everything has the proper appearance.

FIGURE 13-11   Preview your movie before moving on to burn your disc.

To preview your movie, click the Play button below the right edge of the palette window. This will display the Remote Control window, as shown in Figure 13-8. Use the buttons on the Remote Control to control the playback rather than clicking on the menu buttons with your mouse—this will simulate the remote control that anyone viewing your movie in a set-top DVD player will be using.

> **NOTE** *If you have any unlinked buttons on your menus, you will not be able to select those buttons with the onscreen remote control.*

When you are satisfied that everything checks out correctly during the preview, click the Close button in the upper-right corner of the Remote Control window to close the window so that you can continue.

Burning Your Disc

Now that you have finished creating your movie and have previewed it to make certain that everything is correct, you can move on to the final step of burning your disc. In Sonic DVDit!, you have the option of creating three basic types of output:

- You can burn the disc directly. In most cases this will be the option that you will use.

- You can create a disc folder on your hard drive. This creates a set of folders that have the proper layout for a finished disc. You can use this option if you want to do the actual burn at a later time.

- You can create a disc image file on your hard drive. This is similar to the second option, but does not produce the DVD folder structure on your hard drive, so it cannot be tested without first burning the disc. You are unlikely to use this option very often (if at all).

For our example, we will assume that you want to use the first option and create your disc immediately. In addition, we'll assume that you are creating a DVD rather than a CD. To burn your disc, follow these steps:

1. Select Build/Make DVD Disc from the Sonic DVDit! menu.

2. If you see a message like the one shown next, click the "Convert files using current settings" button to convert the files to the proper format.

DVDit!

This project contains files that are not compatible with the DVD Video format. Sonic DVDit! can convert these files for you, so that you can create a DVD. Your original files will remain on your hard disk, so that you can use them again.

[View or modify conversion settings]   [Convert files using current settings]   [Return to Project]

3. When you see the Make a DVD Disc dialog box, as shown next, make certain that Current Project is selected in the Source drop-down list box. You can also choose to use an existing set of DVD folders or a disc image.

4. Select the Include DVD Player checkbox if you want to include DVD player software on the disc so that it can be played in PCs that don't already have DVD player software installed.

5. Select your recordable DVD drive from the drop-down Recorder list box. You can also click the Search button to find the drive automatically.

6. Enter the number of copies to create in the Number of copies text box. If you intend to make multiple copies of your disc, the process will take less time if you make all of the copies at once.

7. Choose Test to test the burn process if you only want to simulate the burn. This will enable you to make certain there are no problems with your DVD recorder. Choose "Test and create disc" to run the test and then create the disc if it appears there are no errors. Choose the Create disc option to skip the test before the burn. This is the fastest option, but until you've made certain that your hardware is functioning properly, it is also the option that is most likely to waste an expensive recordable disc.

8. Make certain that you have inserted a blank, recordable disc in the drive.

9. Click OK to begin burning the disc. It's a good idea to allow this process to have exclusive use of your system to prevent any errors.

In this chapter you have seen how to use Sonic DVDit! to take video files that you have created in another application and use them to build a DVD. You saw that Sonic DVDit! provides considerable flexibility in its main focus area, creating fancy DVD menus.

In the next chapter, we'll move on to creating a complete movie in one of the two mainstream digital video editors we're covering in this book, Pinnacle Studio. Although you have seen a number of examples of using Pinnacle Studio earlier in the book, in the next chapter we'll actually go from capturing video through editing your movie and then on to creating your final output.

CHAPTER 14

Creating a Movie in Pinnacle Studio

There is no doubt that Pinnacle Studio provides far more power and flexibility than any of the other digital video editors we've looked at so far in this part of the book. This program provides you with the ability to create your movies the way you want without holding you back by limiting you to a few predefined templates. Pinnacle Studio also provides you with a full range of tools for each part of the movie creation process. It starts by offering you complete video capture capabilities, and then it moves on to tools for every part of the video editing process. When you have finished developing your movie, Pinnacle Studio enables you to select from many very useful output options that go far beyond simply burning a disc.

If you are considering upgrading your current digital video editing capabilities by purchasing a new program such as Pinnacle Studio, it is important to know that Pinnacle offers several different versions of Studio. These different versions all include the same basic digital video editor, but they differ in the additional items that are included in the package. For example, you can buy a version that does not include any capture hardware if your PC is already equipped with the capture hardware to match your camcorder. You can also buy versions that include analog video capture cards as well as ones that provide an IEEE-1394 (FireWire) digital connection for your DV camcorder. In addition, you have the choice of different versions of Hollywood FX (for fancy transition effects). In other words, you should be able to find exactly the version of Pinnacle Studio to meet your needs. See the Pinnacle Web site (www.pinnaclesys.com) for more information about the available versions.

TIP	*No matter which version of Pinnacle Studio you install, be sure to agree whenever the program offers to check for updates. Pinnacle provides free program updates from time to time, and these often include new features or correct problems in the existing version of the program.*

Capturing Your Video

The first part of any digital video movie creation process is capturing your video content. If you are using a program such as Pinnacle Studio, you don't have to go to another source to capture that video, since Pinnacle Studio is able to capture video from both digital and analog sources (as long as your PC is equipped with the proper capture hardware, of course). You can also import most existing video files and use them in your Pinnacle Studio movie project.

In addition to its video capture capabilities, Pinnacle Studio does an excellent job of scene detection, which can save you a lot of manual editing later in the process. If you use a DV camcorder, Pinnacle Studio can use the recorded time code to automatically detect the scenes, but you have several scene detection options no matter what video source you are using.

Once you have opened Pinnacle Studio, click the Capture tab, as shown in Figure 14-1. This will enable you to choose your capture options and then to begin capturing your video.

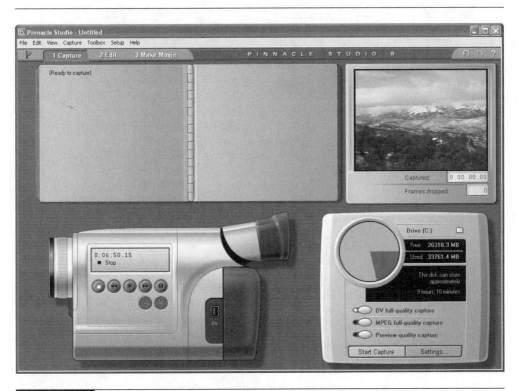

 Use the Capture tab to capture your video content in Pinnacle Studio.

Choosing Your Capture Settings

Before you begin capturing your video, you may want to verify that you have
selected the best set of capture options to suit your project. To do so, follow
these steps:

1. Click the Settings button on the Capture tab to display the Pinnacle Studio
   Setup Options dialog box.

2. In the Pinnacle Studio Setup Options dialog box, click the Capture source
   tab, as shown here:

3. Select the proper source from the drop-down Video list box. If your
   DV camcorder is not listed, make certain it is properly connected and
   powered on.

4. If you are using an analog video source, choose the proper audio input
   from the Audio drop-down list box. (Selecting the DV camcorder as the
   video source automatically selects the DV camcorder as the audio source.)

5. Make certain that the Capture preview checkbox is selected so that you will be able to view the video source during the capture.

6. Choose the scene detection option that best suits your needs. If you use a DV camcorder, the "Automatic based on shooting time and date" option is best since it uses the recorded time code to detect new scenes.

7. If this is the first time you are capturing video, you may want to click the Test Data Rate button to make certain that your connections and your PC are fast enough to handle the incoming video data.

8. Click the Capture format tab, as shown here:

9. If you are using a DV camcorder, make certain that the DV option is selected in the drop-down Presets list box to ensure the highest possible video capture quality. If you are using an analog source, you may want to choose MPEG and then select further options on this tab that suit your project. (Click the Help button for information on the remaining options.)

10. Click OK to close the dialog box. You can choose additional settings on the Edit and the CD and Voice-over tabs later if necessary.

Capturing Video from Your DV Camcorder

Capturing video from your DV camcorder in Pinnacle Studio couldn't be much simpler. As Figure 14-1 showed, you have onscreen controls you can use to advance, rewind, and play the tape in your DV camcorder. Once you have the tape at the position where you wish to begin the capture, click the Start Capture button. This will display the Capture Video dialog box, as shown here:

Capture Video ☒

Enter a name for this capture:

Video 2

Stop capturing after:

540   minutes   4   seconds

Once you start capturing, you can stop at any time, by clicking Stop Capture or pressing Esc.

Start Capture   Cancel

Enter a name for the capture file. If you like, you can also specify a maximum capture length using the minutes and seconds text boxes. Click the Start Capture button in the dialog box to begin the capture. As Pinnacle Studio captures the video, it will use the scene detection method that you specified to determine where to break the video into scenes. Each new scene will appear in the album as it is captured.

> **TIP** *If you use a DV camcorder, Pinnacle Studio will automatically stop capturing when it reaches the end of the recorded section of the tape.*

When you have finished capturing your video, click the Stop Capture button. Then click the Edit tab to continue working with your movie.

Importing Video Files

Sometimes you may want to include in a movie a video that is on a different tape or is a video file that you captured earlier. You can easily use video files that were not a part of the current capture session, but you'll need to know one little tricky detail in order to do so.

The video scenes album in Pinnacle Studio shows the scenes that are a part of a single video capture file. If you examine the Pinnacle Studio menus, you won't find a command that enables you to import additional video into your project. Even so, you aren't limited to using only those scenes that are included in the current capture file. To use scenes from additional video files, however, you'll have to open those files—which automatically closes the current video capture file. Fortunately, this does not affect the scenes that you have already added to the timeline (or to the storyboard). So, the solution is pretty simple. You just have to add scenes to your movie from your capture file, then open the next file and add your scenes from it, and continue until you have added all of the scenes that you want from each of the files. You can even reopen a file if there are more scenes you want to add later.

You open video files in Pinnacle Studio by selecting them from the drop-down list box at the top of the left page of the video scenes album. Here, I've clicked on the down arrow at the right side of the list box to drop the list down so that I can choose which video file I want to open:

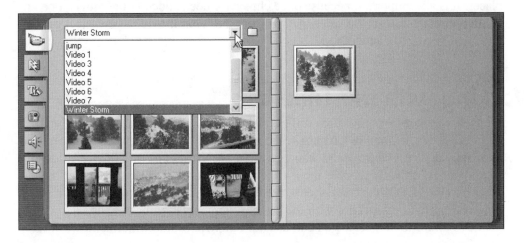

If necessary, you can choose video files from a different folder by clicking the folder icon just to the right of the list box.

Using the Video Scenes Album

As I've mentioned, the video scenes album holds the video clips that you have captured or imported. Only one video file can be open in the video scenes album at a time, but you can simply open and close the video files as needed.

The video scenes album shows an individual thumbnail for each scene. That thumbnail is the first frame of the video clip. As you add a scene to the movie,

Pinnacle Studio adds a small checkmark to the thumbnail in the video scenes album to help you remember which scenes you have already used (this is a feature you'll come to appreciate once you begin creating large, complicated movies where it simply isn't possible to see the entire timeline or storyboard at one time). There is nothing to prevent you from using the same scene more than once, however. For example, you might reuse a scene to create a flashback type of effect.

In many cases you will probably have more scenes than can fit onto the two visible pages of the video scenes album. If so, you can navigate to the other pages of the album by using the forward and back arrows that appear when the album has multiple pages.

It's important to remember that the video scenes album is simply a convenient tool for adding clips to your movies. Once you have dragged a movie clip from the album and dropped it into your movie, the copy of the clip that is contained in your movie can be edited independently of the version that is in the album. This is important because it means that you can use different versions of the same clip in different places. It is also important because once clips have been added to the movie, opening a different video file so you can add clips from it to your movie has no affect on the clips you've already added.

Working with the Timeline

Pinnacle Studio offers you three different options for viewing the layout of your movies. As you learned in Chapter 4, the storyboard, the list view, and the timeline are simply different views of the sequence of the scenes, menus, and titles. (Although we did not specifically discuss the list view, it is essentially just a hybrid of the other two views.) Of the three views, the timeline is by far the most intuitive and powerful option for working with the various elements in your movies.

Let's take a look at how you can use the Pinnacle Studio timeline to work with the scenes in your movies.

Adding Scenes

At the most basic level a movie is a series of scenes that are viewed sequentially. Although it is certainly true that you can use the navigational controls that are associated with menus to modify the flow of movies that you create for disc-based output, the flow of the movie's scenes is normally a linear one from the beginning to the end of the movie. When you add scenes to the timeline, those scenes will appear when the movie is played in the same order that they appear on the timeline.

You add scenes to the timeline by dragging-and-dropping them from the video scenes album onto the timeline. In Figure 14-2, I have added a number of scenes to the timeline. In fact, if you look carefully, you can see that Pinnacle Studio is indicating that six of the scenes have been used (look for the checkmark in the upper-right corner of the thumbnails in the video scenes album). There are another four scenes that I have decided not to use (the ones without a checkmark on their thumbnail).

Figure 14-2 has a number of interesting items to notice. First, the second scene in the movie is displayed with a colored background (it is blue, but you can't tell that from the figure, since it is printed in black and white). This colored background indicates that the scene has been selected. Any editing you may do always affects the currently selected object, so it is important to recognize what has been selected at any time.

Next, although only four scenes appear on the timeline, you can tell that the movie contains additional scenes beyond what is currently visible. The checkmarks

FIGURE 14-2   A number of scenes have been added to the timeline in this movie.

in the video scene album thumbnails are one important clue, and the position of the scrollbar at the far left of its range (at the bottom of the Pinnacle Studio window) is another. (The scrollbar is not a good indicator of the number of scenes, of course, since it's impossible to tell from the current view how long the off-screen scenes might be.) Clearly, though, you can view additional scenes in the movie by dragging the scrollbar to the right.

Finally, consider the position of the timeline scrubber in Figure 14-2. It is currently sitting at the start of scene 2, which means that the viewer window is showing the view of the movie at that point in the timeline. If you move the timeline scrubber to a different point on the timeline, you'll see a different frame in the viewer window.

TIP	*Remember that you can also rearrange clips on the timeline by dragging-and-dropping them where you want them to appear. The remaining scenes will automatically adjust their positions to make room for the new clip and to close up any resulting gaps in the timeline.*

Trimming Scenes

There are any number of reasons why you may want to do some trimming of the scenes that you have added to your movies. Perhaps you shook the camera at the beginning or at the ending of the scene when you were trying to make some adjustments. Or possibly someone walked in front of the camera right after you started shooting. It could just be that the scene is too long and boring the way it is. Whatever the reason, you'll find that trimming your scenes is quite easy to do.

Figure 14-3 shows scene 2 being trimmed to cut off about ten seconds from the beginning of the scene. In this case, we're using the video toolbox to do the trimming, since it provides the most precise control over how much is actually trimmed. You can also trim scenes by dragging the edge of the scene directly on the timeline.

NOTE	*If you are trying to produce a video that is a specific length by trimming your video clips, remember that adding transitions between scenes reduces the overall length of the movie by the length of the transitions that you add. You will need to keep this in mind so that you do not trim too much from the scenes and thereby end up with a video that is shorter than you had planned.*

Adjusting the Playback Speed

Another way to adjust the amount of time that a video clip plays is to adjust the playback speed. You can speed up or slow down a scene as necessary to achieve the results you want.

Set a precise starting point here.

Drag this marker to trim the ending of the clip.

Set a precise ending point here.

View the current clip duration here.

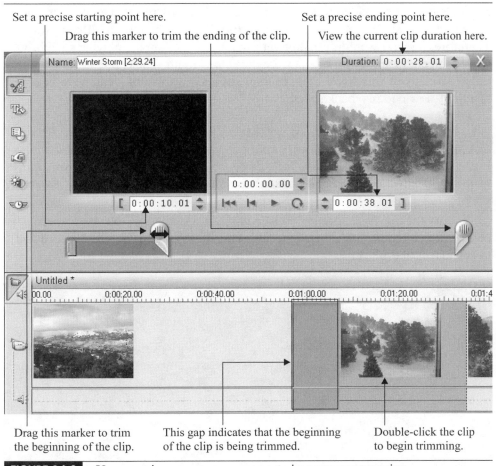

Name: Winter Storm [2:29.24]   Duration: 0:00:28.01

0:00:00.00

[0:00:10.01   ◄◄ ◄ ► ↻   0:00:38.01]

Untitled *

0:00.00   0:00:20.00   0:00:40.00   0:01:00.00   0:01:20.00   0:01:4

Drag this marker to trim the beginning of the clip.

This gap indicates that the beginning of the clip is being trimmed.

Double-click the clip to begin trimming.

FIGURE 14-3   You can trim scenes as necessary to improve your movies.

One short movie I recently produced provided an excellent example of how adjusting the playback speed could be very useful in adjusting the length of the playback. I had set up my camcorder to shoot a time-lapse movie showing a stormy day as it appeared from my office window. On that particular day the cloud patterns and the snow squalls in the mountains outside my window produced quite an impressive show. I programmed my Sony DCR-TRV27 DV camcorder to take a few frames of video at regular intervals throughout the day. When the day was done, the resulting video was quite impressive.

In creating my movie, I decided that the overall effect would be greatly enhanced by adding a rendition of *Winter* from Vivaldi's *Four Seasons,* so I ripped a copy of the piece from an audio CD I had sitting around. Now I had just one problem— at normal playback speed the video went on for about a minute past the end of the music. But since the video was showing events that took place over an entire day in just a few minutes, the playback speed didn't really match real life anyway. Therefore, it really wouldn't matter very much if I sped up the playback slightly so that the audio and video tracks were the same length. The end result was quite impressive, and it was also a great example of just how useful it can be to be able to adjust the video playback speed.

NOTE *Remember that if you play a video clip at other than normal speed, the recorded audio track will not play. If you want to use the recorded audio track you will need to separate the video and audio tracks before you adjust the playback speed. We'll discuss audio tracks a bit later in this chapter.*

To vary the playback speed of a video clip, you use the Vary playback speed tab of the video toolbox shown here (the tab's two sliders are for playback speed and to create a strobe-like effect):

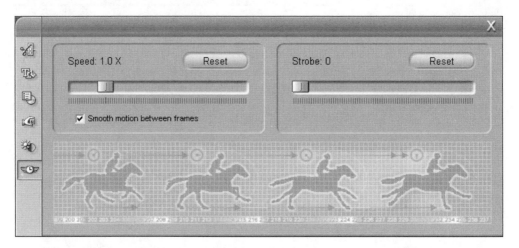

Here is what you need to do in order to adjust the playback speed:

1. Select the video clip that you want to adjust.

2. Click the video toolbox icon (the icon that looks like a camcorder) at the upper-left corner of the timeline to open the toolbox, if it is not already open.

3. Click the Vary playback speed button (the lowest button along the left side of the video toolbox) to display the Vary playback speed tab.

4. Drag the Speed slider to the right to speed up the playback or to the left to slow it down.

5. Make certain that the "Smooth motion between frames" checkbox is selected if you are slowing the playback to less than normal speed. This will help make the video look a bit less jerky during playback.

6. To produce a strobe effect where frames are repeated to simulate stop-action photography, drag the Strobe slider to the right.

7. Click the Play button below the viewer window to preview the results.

Adjusting Scene Properties

If you find that a scene doesn't look quite right because it is too dark, because it is too washed out, or because the color balance is not quite correct, you can use the "Adjust color or add visual effects" tab (shown here) to make some corrections:

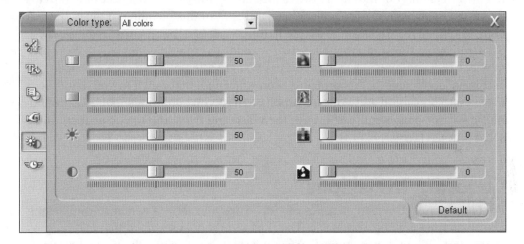

This tab also enables you to adjust some additional properties to create some interesting special effects (move the mouse pointer over each of the sliders to see a description of the property that the slider will adjust).

The "Adjust color or add visual effects" tab of the video toolbox is displayed when you click the second button from the bottom at the left side of the video toolbox. In addition to experimenting with the sliders, you may want to check out the options in the drop-down Color type list box. You can use these options to change a scene to black and white, sepia, or a single color if you want to give the scene a special appearance.

When you are finished adjusting the scene properties, click the video toolbox icon again to close the video toolbox. At this point, you should have a movie with several scenes on the timeline so that we can continue building your movie.

Using Transitions

Transitions can be a very useful tool for helping you get across the story that you want to tell with your movie. Unfortunately, transitions can also be an unnecessary distraction if you overuse them or if you use ones that simply don't match the style of your movie.

Pinnacle Studio provides you with an extremely broad selection of transitions, so with a little care in your choices you should be able to locate transitions that enhance your movies. Depending on the version of Pinnacle Studio that you have installed on your system, you may even have the option to modify the transitions (see Chapter 5 for more information on this subject).

Transitions work by combining the ending of one clip with the beginning of another. Each particular style of transition produces slightly different visual effects, but the effect on your movie is always the same—the overall length of the movie is decreased by the length of the transition as a result of this blending of the two video clips. (I realize that this can be a little hard to visualize, but it's true regardless.) So if you add a transition that takes two seconds to complete, the length of your movie is reduced by two seconds. Unless you are attempting to create a video that is some specific length, this probably won't mean all that much to you, but it does mean that you should add your transitions before you add things like background music (since you generally will want the video and the audio to end at about the same time).

Choosing Transitions

For our example, I'm using a series of video clips that show a winter storm blowing fiercely outside my office windows. The clips were shot in several different directions, so it seems as though transitions that move across the screen in the general direction that the snow is blowing in the new scene might work well.

There are three types of transitions that fit this definition: *wipes, slides,* and *pushes.* Considering the characteristics of the three, a push transition would be a good choice, since it will do the best job of conveying the power of the storm to push things around.

Your own movie will probably entail different considerations that will affect the types of transitions you should use. When in doubt, experiment with the different styles—don't be afraid to decide that none of the transitions is an improvement over a simple direct cut from one scene to the next.

Adding Transitions

Once you have decided which transitions will fit into your movies, you're ready to begin adding them. To do so, follow these steps:

1. Make certain that both the video toolbox and the audio toolbox are closed. (If either of the buttons at the upper-left corner of the timeline are depressed, one of these toolboxes is open.)

2. Click the Show transitions tab (the second from the top on the left side of the album pages) to open the transitions album, as shown here:

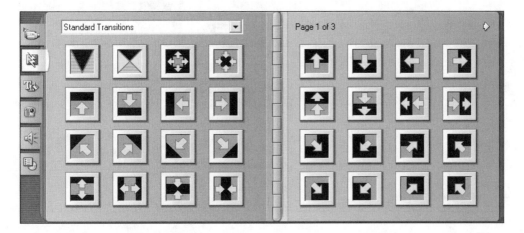

3. Click the transition you want to use and then watch the viewer window to preview the transition. Remember that the drop-down list box at the top of the left page of the transitions album lists additional transition categories that you can choose to use in your movies.

4. Drag-and-drop the transition from the transitions album onto the timeline. Be sure to drop it between the two scenes that you want the transition to affect.

5. Click the Play button below the viewer window to preview the transition using your two video clips.

6. To change the length of the transition precisely, double-click the transition in the timeline to display the video toolbox, as shown here:

7. Use the Duration box at the upper-right of the video toolbox to adjust the length of the transition.

8. Click the Close button on the video toolbox to close it.

9. Continue adding transitions between scenes until you have added as many as you want to use in your movie.

> **TIP**   *You can also drag the ends of the transition inwards or outwards to adjust the length of the transition. This method is not nearly as precise as using the video toolbox, but it may be precise enough for your needs (and it is a little faster than using the video toolbox).*

Be sure to use the File/Save command on the Pinnacle Studio menu from time to time to save your movie project as you work on it.

Adding Titles

Next, we'll add some titles to our movie. It's pretty clear that titles add a little fancy touch to movies, and they often serve to inform the audience what the movie is about, who stars in it, and (most importantly) who created it.

Pinnacle Studio enables you to add two different types of titles. There are the standalone titles that can appear over a still or moving background, and there are the title overlays that appear in front of scenes from your movie. You can use whichever type of title best suits a particular point in your movie—and you're free to use both types, if you like.

> **TIP**
>
> *One important difference between standalone titles and title overlays is their affect on the length of the movie. Adding a standalone title increases the length of the movie by the duration of the title, while adding a title overlay has no affect on the length of the movie.*

You add standalone titles to the main video track on the Pinnacle Studio timeline, and you add title overlays to the title track. In either case you have several different ways to add a title. Let's take a quick look at two of them.

Using Title Templates

The title album includes 36 pre-made title templates that you can easily use to add either a standalone title or a title overlay. This shows a sample of some of those pre-made titles:

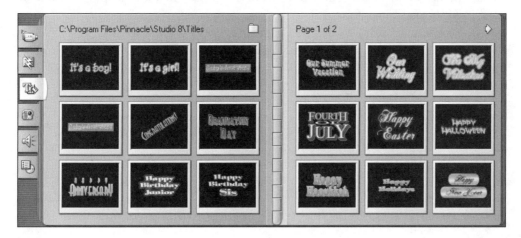

The title templates appear in the title album (the third tab in the album).

To add one of the pre-made titles to your movie, you can drag-and-drop the title from the title album onto the timeline. To create a standalone title, drop the title onto the main video track. To create a title overlay, drop the title onto the title track. Try to drop the title where you want it to appear in the movie, but if you're a little off don't worry, because you can always fine-tune the location by dragging-and-dropping.

Although the pre-made title templates provide a quick way to add some simple titles to your movies, they're probably not quite what you'll want in many cases. Some of them even include placeholders that say "Your Name" and are clearly meant to be modified rather than simply being used as is. To modify them, you'll need to use the title editor, which also happens to be the tool you'll use to create your own titles from scratch. Let's take a look at creating your own titles using the title editor.

Creating Your Own Titles

In most cases, you'll probably want to create your own titles from scratch using the Pinnacle Studio title editor. Doing so is almost as easy as using one of the pre-made title templates and ensures that you'll get exactly the title that you want.

To create your own titles from scratch using the title editor, follow these steps:

1. Right-click the point in the timeline where you want to add a title. Remember to right-click in the main video track if you want to create a standalone title or in the title track if you want to create a title overlay.

2. Select Go to Title/Menu Editor from the pop-up menu to display the title editor, as shown here:

3. Select a text style from the list along the right side of the text editor window by clicking the style that you want to use.

4. Drag out a text box in the viewer window where you want your text to appear. Remember to stay inside the red dotted line that indicates the TV safe zone.

5. To change the text attributes (such as the typeface, alignment, or size), use the text attribute controls above the right side of the viewer window.

6. Enter your text in the text box.

7. To make your title roll up, select the Roll button (above the left edge of the viewer window). To make your title crawl from right to left across the screen, select the Crawl button. When you select either of these options, the other title style options will automatically be deselected.

8. If you want to adjust the face, edge, or shadow of the text, click the Custom tab (near the upper right of the text editor window) and then use the sliders to make your adjustments. Note, however, that you may need to select a different typeface if you want to adjust the edge or the shadow, since some typefaces do not allow this type of modification.

9. If you are creating a standalone title, use the Backgrounds or the Pictures button to choose a background image for the title. Remember that standalone titles are an addition to the main video track, so they don't appear over a background image unless you specifically add one.

10. To make the title appear for a specific length of time, adjust the length using the Duration control in the upper-right corner of the title editor window. You can also adjust the duration once the title is on the timeline by dragging the edges of the title clip.

11. Click the OK button when you are finished creating your title to close the title editor and to place your title onto the timeline.

TIP	*You can also use transition effects between title clips. This is a very handy technique for producing titles that scroll onto the screen, pause for a short time so that they can be read, and then scroll off the screen to be replaced by another title. If you place a transition at the beginning or at the ending of a sequence of title clips, the transition effect will make the first (or last) title scroll onto or off of the screen without another title being affected.*

Figure 14-4 shows the movie with a title overlay and a transition added to the timeline. In this case, I am previewing the title.

Adding Sounds

When you capture a video from your camcorder, Pinnacle Studio automatically adds the sounds that were recorded along with the video. If you later add the video clip to the timeline, the recorded audio and recorded video tracks are synchronized so that the audio and video playback occur together at the same timing as when they were recorded. As a result, if you see someone talking in the video, their lips will be moving in sync with their speech.

Sometimes, however, the recorded audio track doesn't quite do everything that you might want. For example, you might want to add some narration, some sound effects, or even some background music. In some cases, you might even want to dispense with the recorded audio track completely and replace it with other audio tracks. Let's have a look at each of these options to see how you can apply them in Pinnacle Studio.

A title overlay added to the title track

A transition added between scenes

Preview the title here.

FIGURE 14-4   A title and a transition have both been added to this movie.

Separating the Recorded Audio Track from the Video

If you want to work with the recorded audio track separately from the main video track, you first need to lock the video track, as shown in Figure 14-5. A locked track is unaffected by any changes you make to other tracks in the timeline, and locked tracks cannot be accidentally selected.

In Figure 14-5, I have locked the main video track and then selected the recorded audio tracks that were associated with the first three video clips. (Remember to hold down CTRL as you select additional tracks to add them to the selection—you can also hold down SHIFT to select a contiguous range.) Once you have selected all of the recorded audio tracks you can remove them by clicking the trashcan icon.

Click here to lock the video track.   Click here to delete the selected clips.

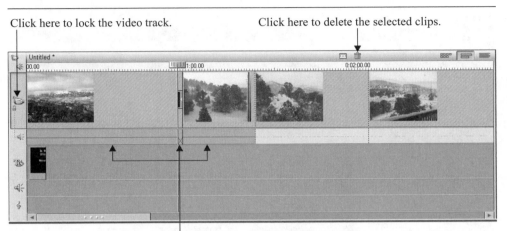

Once the video track is locked, click the recorded
audio track to select it separately from the video track.

FIGURE 14-5   Here I'm getting ready to remove the recorded audio track.

TIP   *If you aren't sure that you want to permanently remove the recorded audio
track, you can also use the audio toolbox to reduce its volume level to the
lowest possible level. This may not always completely remove any especially
loud sounds on the track, however.*

Adding Music

Next we'll add some background music to our movie. You can add music from
an audio CD, but in this case we'll use the SmartSound feature in Pinnacle Studio
to add some background music that is exactly the length of the movie. (And as
I mentioned in Chapter 6, using SmartSound background music bypasses any
copyright issues, so we can use the movie any way we want without worrying
about legal problems.)

To add the background music, follow these steps:

1. Click the button at the left side of the main video track to unlock the track.
   You need to unlock the track so that you can select the video clips on
   the track.

2. Click the Audio Toolbox button (the button that looks like a speaker at the
   upper-left corner of the timeline) to open the audio toolbox.

3. Click the "Create background music automatically" tab in the audio toolbox to display the SmartSound options, as shown here:

Name:		Duration: 0 : 00 : 05 . 00 ‡ X

Style	Song	Version
Classical	Four Seasons	Parlor Piano
Country/Folk	Mozart Night	High Twinkle
Jazz/Fusion	Piano Sonata	Regal Piano
New Age/Easy	Queen of Sheba	
Orchestral		
Pop/Dance		
Rock		
Specialty		
All		

Preview	Add to Movie	SmartSound® ...

4. Choose the style, song, and version options that suit your tastes.

5. Click the Preview button to hear a short sample of your selected music. Remember that this sample will play for only a few seconds, but the music you add to the timeline will play for the entire length of the movie.

6. Select all of the video clips. The easiest way to do this is to press CTRL-A or select Edit/Select All from the Pinnacle Studio menu. By selecting all of the video clips, you will ensure that the automatically generated background music is the same length as the movie.

7. Click the Add to Movie button to add the background music track to the timeline.

8. Click the Play button below the viewer window to view your movie while you listen to your new background music track.

9. Select File/Save from the Pinnacle Studio menu to save your work.

TIP
If the background music track seems too loud, click the second button from the top of the audio toolbox to display the volume controls so that you can adjust the volume level. The background music track is controlled by the right-most volume control.

Adding Voice-Overs

We have all heard voice-over audio tracks in movies. An announcer might explain that the video you are seeing depicts the long journey of the Donner party crossing the 40-Mile Desert on their way to the Sierras, for example. This type of narration can be a very effective method of informing your audience about something they might not otherwise pick up from simply watching your video.

Pinnacle Studio provides a third audio track in the timeline that you can use to add voice-over narration or for sound effects. (Actually, Pinnacle Studio allows you to place any type of audio on any of the three audio tracks, but simple tools are available if you add the default track types to the timeline.)

To add narration to your movie, you use the "Record a voice-over narration" tab of the audio toolbox, as shown here:

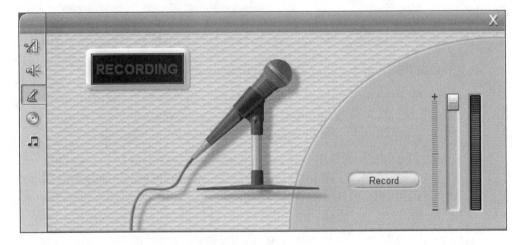

This part of the audio toolbox records a narration track using the microphone that you have attached to your PC.

To record a voice-over narration, follow these steps:

1. Move the timeline scrubber to the point in your movie where you would like the narration to begin. You can move the voice-over clip after it has been recorded, of course, but it will probably be less confusing if you start in the correct place.

2. If the audio toolbox is not open, click the audio toolbox button at the upper-left corner of the timeline to open it.

3. Make certain that your microphone is properly positioned (I find that headset microphones generally produce the best results when recording a voice track on a computer).

4. Click the Record button to begin a three-second countdown (watch the Recording indicator in the audio toolbox).

5. When the countdown finishes, begin speaking in your normal tone of voice.

6. Watch the recording level indicator along the right side of the audio toolbox. Use the volume control slider to keep the indicator within the yellow band as much as possible.

7. Click the Stop button when you are finished recording your voice-over track.

8. When the standby indicator disappears, move the timeline scrubber back to the beginning of the voice-over track.

9. Click the Play button under the viewer window to listen to the narration as you watch the video playback.

10. If you made any mistakes or are otherwise unhappy with the results, click the voice-over clip to select it and then click the trashcan icon to delete the clip. Then start over and record the narration again.

You'll probably find that it is helpful to write out a script and practice the narration while watching the video before you record the voice-over track. Even a quick run through will likely make you more comfortable and produce better results. At the very least, I would suggest that you prepare some notes so that you don't forget any of the important points of your narration during the recording session.

TIP	*Remember that you can also use a voice-over track to create a humorous effect, such as very poorly synched dialog.*

Adding Sound Effects

Sound effects are another type of audio that you might want to add to your movies. Sound effects can produce an ambience that might otherwise be difficult to achieve because of the difficulty in accurately recording certain types of sounds. For example, in the following illustration you'll see I've selected the ColdWind sound effect from the Pinnacle Studio sound effects album to add to my winter storm movie.

Unless I had wanted to stand outside in a blizzard during the filming, it would have been very difficult to capture this type of sound effect myself.

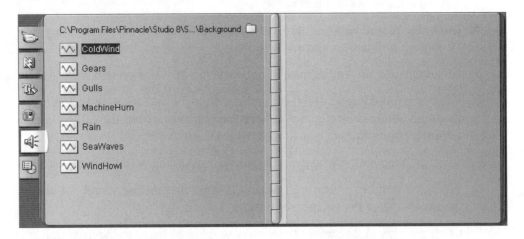

As I mentioned earlier, the voice-over track and the sound effect track are the same track in the timeline. Unless you move one of them to a different audio track, you cannot have overlapping narration and sound effects. This is not a problem in the case of the winter storm movie, however, since the audio track that is normally used for the recorded audio is empty because I sent that track to the trashcan.

> **TIP** *You can have both narration and sound effects on the same track as long as you don't try to make two audio clips overlap on one track.*

You add sound effects using the sound effects album rather than the audio toolbox. This may seem a little confusing at first until you remember that sound effects are simply recorded sound files that you are adding to your movie. Whenever you are adding preexisting content to a movie, you use one of the albums.

When you first open the sound effects album, you will see a limited set of sound effects from which to choose. In reality, though, Pinnacle Studio has quite a few sound effects that are stored in several different folders. To select additional types of sound effects, click the folder icon in the sound effects album to display the Open dialog, box shown here:

Next, click the Up One Level icon (the folder with the up arrow to the right of the drop-down Look in list box) to see the categories. Open a folder to see what it contains, then click the Open button to display its contents in the sound effects album.

Don't forget that you can use the audio toolbox to trim a sound effects clip that you have added to your movie. You can also repeat the same sound effects clip if you want the sound effect to last for a longer period of time.

Creating Menus

Menus enable viewers to navigate through your movies. You may not have thought very much about menus, but they provide many capabilities that simply don't exist in a linear medium such as videotape. Consider the following example and you'll have a better idea what I mean.

Suppose that you have decided to put together a movie that contains footage from your last family gathering. If you simply create a movie that runs straight through from beginning to end without any indication about where a particular sequence might appear, someone would have to sit down and watch the entire video in order to find the clip where Uncle John tripped over the dog and smashed the cake as he fell onto the picnic table. But imagine how much more fun your movie might be to watch if you could just click a button and jump right to the bloopers section of the movie. Sure, when someone has the time to sit through a couple of hours of video they can watch the whole movie, but adding some quick navigation links would probably ensure that your video would be a whole lot more popular (at least with some family members).

Menus appear on the main video track of the timeline, but they are treated somewhat differently from ordinary video clips. Whenever the timeline scrubber is located within a menu clip, the scrubber is within a loop. That is, the scrubber will advance normally during the menu clip, but when the end of the clip is reached the scrubber automatically returns to the beginning of the same menu clip. This process continues indefinitely until someone clicks one of the menu buttons (which causes the scrubber to jump to the timeline location that is associated with the button link).

Adding Menus

You can add menus to your movies using the menus album or by using the title and menu editor. Since we used the editor earlier in this chapter to add some titles, we'll take the easy way out with our menus and add them from the menus album for this example.

To add a menu to your movie using the menus album, follow these steps:

1. If either the video toolbox or the audio toolbox are open, close them so that you can view the albums.

2. Click the Show menus tab (the bottom tab on the left side of the albums) to display the menu templates, as shown here:

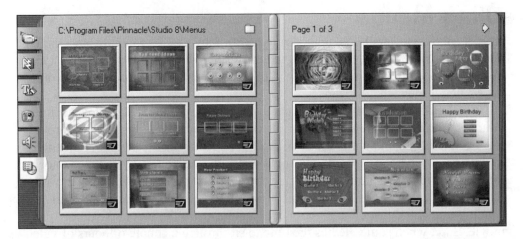

3. Select the menu template that you want to use and then drag-and-drop it onto the main video track on the timeline. You will probably want to drop it at the beginning of the timeline, so that the menu appears as soon as your disc begins playing.

4. Click Yes if the following dialog box appears so that Pinnacle Studio will automatically create links to the scenes (you can delete any links you don't want later). It is possible that this dialog box won't appear if you have selected the "Don't ask me this again" checkbox.

5. Use the viewer window playback controls to test the menu.

<table>
<tr><td>TIP</td><td>Don't forget that you can add music to your menus in much the same way that you added it to video clips.</td></tr>
</table>

Controlling Links

Although using one of the menu templates to create your menus is quite easy, it's entirely possible to end up with too many menu buttons if you use the option to automatically create a link to each scene. If you would like to reduce the number of buttons and eliminate some of the unnecessary links, the process is quite simple. You can also add new links, move links, or even set up your movie so that the menu is redisplayed after specific scenes have finished playing (the menu automatically appears after the entire movie finishes playing).

To edit the links, right-click the menu track that is added to the timeline above the main video track as soon as you add a menu. This will display the pop-up menu shown here:

<table>
<tr><td>Delete</td></tr>
<tr><td>Set Disc Chapter
Set Return to Menu</td></tr>
</table>

You can now choose from the following options:

- **Delete**   Select this option to remove the selected chapter point. This also removes the associated button from the menu.

- **Set Disc Chapter**   Choose this option to add an additional chapter point. This is most useful when you want to add a jump to a point within a scene rather than simply to the beginning of the scene.

- **Set Return to Menu**   Use this option if you want the menu to reappear once the selected scene finishes playing. This makes it possible to create more than one independent movie on a single disc.

More of the fancy menu tricks you can apply in Pinnacle Studio were described in Chapter 8.

Creating the Final Output

Although we could easily continue to play with the many editing options in Pinnacle Studio, it's time to move on and create your final output. Let's begin by looking at your output choices.

Choosing Your Output Medium

One of the features that really sets Pinnacle Studio apart from the basic digital video editors is the number of different output options that it makes available to you. Yes, you can easily create a DVD or VCD, but you can also choose to save your movie on videotape or to create specialized files for several different computer playback options.

NOTE   *Pinnacle Studio maintains full quality in your video and audio tracks throughout the movie creation process right up until you choose to create your final output. This means that you can produce the best quality movies for several different output types from a single movie project. Be sure to save your movie project before you render the final output so that you can easily choose a different output format with no loss of quality.*

When you click the Make Movie option near the top of the Pinnacle Studio window, you'll see the section of the program that is shown here:

Notice that the left side of this section has buttons that enable you to choose the different output options.

If you click the Settings button, you can also choose specific settings for each of the output options using the Pinnacle Studio Setup Options dialog box shown here:

Notice that the tabs along the top of this dialog box roughly conform to the choices in the Make Movie section. The RealVideo and Windows Media output options fall under the Stream setting, and the Share setting creates a video file for uploading to the Web.

You may want to refer to Chapter 9 for more information about the various output options to help you choose the one that best suits your needs.

Burning Your Disc

Once you have selected the proper output options for your movie, you're ready to burn that disc. To do so, start the process by clicking the Create Disc button (this button will display slightly different text if you have selected a different type of output for your movie).

Clicking the Create Disc button starts the rendering process (this is called *transcoding* in some programs). Rendering converts your raw video and audio into the proper type of file for your selected output. For most output formats, rendering also compresses the file so that it takes less space on the disc. Rendering can take considerable time—the amount of time depends on a large number of factors, including the performance of your PC, the size of the movie, the output format that you have selected, and the complexity of your movie. As the rendering takes place, you'll see a couple of progress bars below the viewer window. Don't plan on using your computer for anything else until the rendering and the burning processes have finished.

When the rendering process completes, you'll need to insert a blank disc into your recordable drive. Once you have finished burning the disc, give it a try in your set-top DVD player. You may want to try creating a few more movies before you decide to quit your day job and become a movie mogul, but I'm sure you'll be pleased with the results. (I'm also certain that you'll see some room for improvement with your first few movies.)

In this chapter, I stepped you through the process of creating an entire movie in Pinnacle Studio. Although you have seen examples of how to use this program in earlier chapters, you finally got the opportunity here to take a project all the way through to completion.

In the next chapter, we'll look at our final example of digital video editors as we tackle the task of creating a complete movie in Roxio VideoWave Power Edition. As you will see, it is another very fine product that may well be the perfect tool for your needs.

CHAPTER 15

Creating a Movie in Roxio VideoWave Power Edition

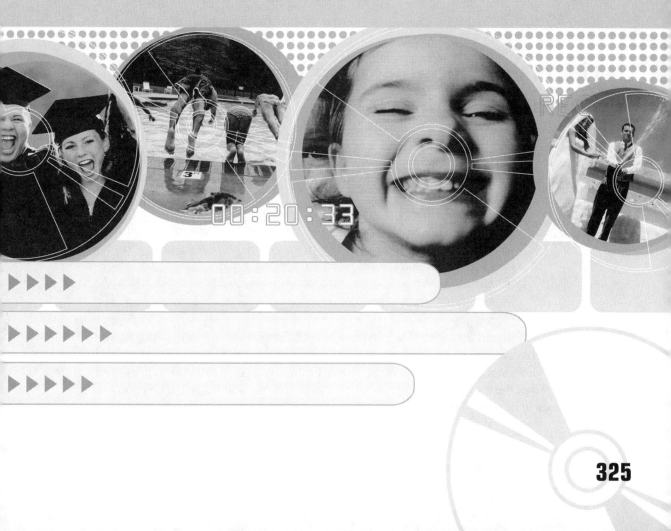

▶▶▶▶▶

▶▶▶▶▶▶▶

▶▶▶▶▶

We are finally down to the last digital video editor that we will be looking at, Roxio VideoWave Power Edition. This is an extremely versatile program that offers many features that you won't find in the entry-level editors. Roxio VideoWave Power Edition has a deceptively simple appearance that hides a whole toolbox full of very powerful tools.

Roxio VideoWave Power Edition is aimed at an audience that does not need (or expect) nearly the level of hand-holding that you find in the entry-level editors. In other words, it is intended for someone—like you, a reader of this book—who has an understanding of the various steps that go into producing a DVD. This does not make Roxio VideoWave Power Edition harder to use; it simply means that the program provides you with a lot of flexibility, so that you can do most things your way without a bunch of artificial barriers to hold you back. If you want to create DVD movies that you can be proud of, you are bound to like Roxio VideoWave Power Edition.

Understanding the Roxio VideoWave Power Edition Screen

Roxio VideoWave Power Edition has very little resemblance to any other Windows-based program that you have ever encountered. As Figure 15-1 shows, the Roxio VideoWave Power Edition main screen has a unique layout that is set up specifically with the video editing process in mind.

NOTE *When Roxio VideoWave Power Edition opens, it automatically maximizes itself to provide you with the greatest possible workspace. You cannot reduce the size of the Roxio VideoWave Power Edition window—this may give you a clue that digital video editing takes a lot of your computer's resources, so maybe you should plan on not working on other types of projects at the same time.*

It is important to familiarize yourself with the various sections of the Roxio VideoWave Power Edition window and also with the controls that appear in the window. Since Roxio VideoWave Power Edition does not lead you step-by-step through the movie creation process, you need to understand what each control does and when you will need to use it.

New production   Cutting room
   Open production   Darkroom   Transitions
    Save   Special effects   Video mixer
     Undo   Text animator   Storyline

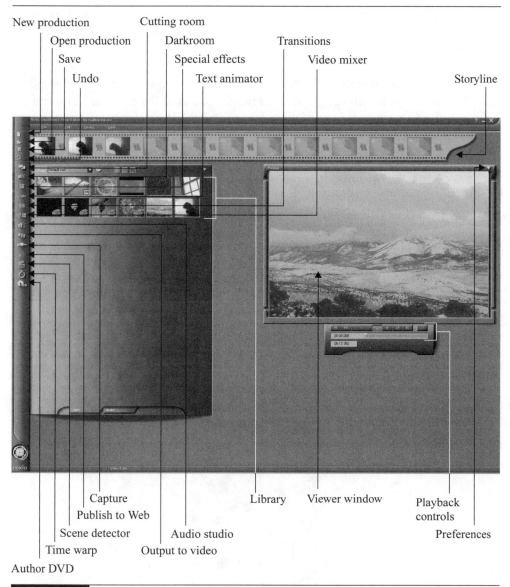

          Capture   Library   Viewer window   Playback
       Publish to Web   controls
     Scene detector   Audio studio   Preferences
   Time warp   Output to video
Author DVD

FIGURE 15-1   Roxio VideoWave Power Edition looks quite different from other
programs.

In most cases the Roxio VideoWave Power Edition window is divided between the sections that you see in Figure 15-1. Let's take a close look at the purpose of the various areas in this window:

■ The toolbar along the left side of the Roxio VideoWave Power Edition window provides you with access to all of the functions of the program. You can learn the purpose of any of the buttons by simply holding the mouse pointer over the button for a few seconds until a small caption displays.

■ The storyline near the top of the Roxio VideoWave Power Edition window is where you assemble your movie. You drag-and-drop video clips or image files onto the windows on the storyline. If you drop a clip onto one of the windows that already contains a clip, the existing clips will move to the right to make room for the new clip. When you want to add transitions between two scenes, you drag-and-drop the transition into the area between the two clips.

■ The library is the area in the left half of the Roxio VideoWave Power Edition window where your captured or imported movie clips appear so that you can use them in your movies. If you right-click one of the clips in the library, you will open the pop-up menu shown here (although the Show Scenes option will appear only if you have used the scene detection option to split the clip into scenes).

■   The viewer window in the right half of the Roxio VideoWave Power Edition window enables you to preview video clips or your movie. It is also the area where you will use the various tools to edit your videos. When you are working with one of the editing tools, the area below the viewer window will display the tools and controls you use for the editing process.

TIP   *Eventually, you will probably add more clips to the storyline than you can see on a single screen. You can scroll the storyline by moving the mouse pointer over the sprocket holes on the storyline and then dragging the storyline left or right when the mouse pointer changes to a hand.*

Adding Video to a Project

In any movie project, you need to start by adding the video content that you want to include. In Roxio VideoWave Power Edition you can use both digital and analog video sources (if your PC is equipped with the proper capture hardware, of course) as well as existing video files that you have already captured. In addition, you can add still images and audio files to your projects.

When you add files to a Roxio VideoWave Power Edition project, those files appear in the library area of the window. Once the files have been added to the library, you can add them to your movie. It is also possible to tell Roxio VideoWave Power Edition to automatically add captured video to the storyline, but in most cases selecting the video that you want to use from the library is a better way to build your movie, since doing so gives you more control over the final movie layout.

Capturing Your Video

Capturing new video from your DV camcorder or from an analog video source will in most cases be the method that you will use to add your video files to your movie project. This is a simple process and one that you will likely have no trouble understanding as soon as you give it a try.

Before you begin capturing video, make certain that your DV camcorder (or other video source) is properly connected to your PC, that it is powered on, and that it is set to the correct mode for transferring video to your computer. This will ensure that Roxio VideoWave Power Edition can properly identify the video source.

When you have your DV camcorder set up and ready to begin the transfer, click the Capture button in the Roxio VideoWave Power Edition toolbar at the left

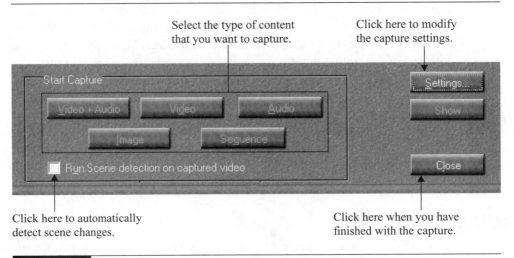

Select the type of content that you want to capture.

Click here to modify the capture settings.

Click here to automatically detect scene changes.

Click here when you have finished with the capture.

FIGURE 15-2 Use these controls to determine what is captured.

side of the screen. You can use the playback controls to verify that the DV camcorder is connected correctly and that the tape is at the beginning of the section that you want to capture.

Setting Your Capture Preferences

There are several options that you can set to control the capture. When you click the Capture button, the controls shown in Figure 15-2 will appear in the lower-right corner of the Roxio VideoWave Power Edition window. You use these controls to set some important options.

In most cases you will want to make certain that the "Run Scene detection on captured video" checkbox is selected. This will tell Roxio VideoWave Power Edition to automatically detect scene changes in the captured video so that you will not have to do so manually later. When Roxio VideoWave Power Edition has run scene detection on a video clip you can choose to add individual scenes to the storyline rather than add the entire clip to the storyline (you can still use the whole clip if you want to, of course).

Next you should click the Settings button to display the Capture Settings dialog box. If you have more than one video source you can select the correct one from this dialog box. In most cases, however, you can ignore the options in this dialog box if you always capture video from the same source. You will want to click the Preferences button in the Capture Settings dialog box to display the Capture Preferences dialog box shown here:

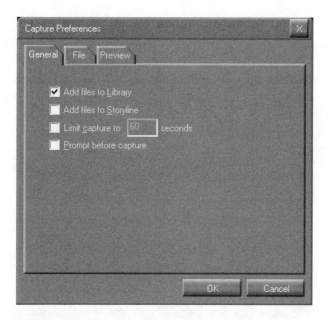

This dialog box enables you to specify where you want the captured files to appear and offers some additional useful options.

Once you have selected the options that you prefer on the General tab of the Capture Preferences dialog box, click the File tab, as shown next. You can use this tab to set the automatically generated file name for the capture files, the location where the files are stored, and the file type for still image captures.

Next, click the Preview tab of the Capture Preferences dialog box, as shown here. On this tab, you can set the size of the viewer window. You'll probably want to set this to match the resolution of your video source, but you may want to set it to a smaller size if your screen resolution is too low (which makes the viewer window too large in proportion to the rest of the Roxio VideoWave Power Edition window). If you have problems capturing video properly, you may find that selecting one or both of the two checkboxes on this tab will resolve the problem.

Close both the Capture Preferences dialog box and the Capture Settings dialog box to return to the Roxio VideoWave Power Edition window. You will use one of the five buttons shown in the Start Capture box (see Figure 15-2) once you begin playing your video to start the capture.

Beginning the Capture

Now that you have your capture preferences set, you are almost ready to begin the capture. First, though, take a look at the new options that have appeared in the playback controls area below the viewer window, as shown in Figure 15-3. These new options provide some additional tools you can use to control the capture.

The playback shuttle control enables you to move quickly forwards or backwards through the tape to find the exact location where you wish to begin

Playback shuttle control

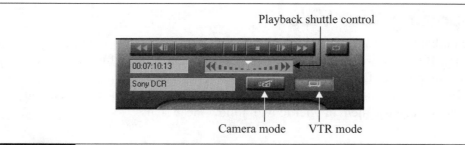

Camera mode   VTR mode

FIGURE 15-3   During capture the playback controls include some additional options.

the capture. You drag the pointer to the right to fast-forward the tape—the further you drag the pointer away from the center, the faster the tape plays. You drag the pointer to the left to play the tape in reverse.

The Camera mode and VTR mode buttons control what is recorded from your camcorder. In VTR mode (Video Tape Recorder), the capture source is the tape that is contained in your camcorder. In Camera mode, you capture a live video feed from your camera's lens. Whichever mode you select must match the settings on the camcorder. On my Sony DV camcorder, VTR mode is called "VCR," while Camera mode is called "Camera."

To capture your video, follow these steps:

1. Use the playback controls to position the tape where you want to begin capturing the video.

2. Click the green Play button.

3. Click the proper button in the Start Capture area. This will be the button that indicates what you want to capture (Video + Audio in most cases).

4. Click the Stop button in the Start Capture area to stop capturing video.

5. When you have completed the capture of all of the scenes that you want to include, click the Stop button in the playback controls to stop the tape.

6. Click the Close button (in the lower-right corner of the Roxio VideoWave Power Edition window) to finish the capture session. This will also add your newly captured video to the library.

Importing Your Video

You probably have some existing video files that you may want to use in your movies. If so, you will find that it is quite easy to import those files into Roxio VideoWave Power Edition.

To import existing video files into the Roxio VideoWave Power Edition library so that you can use them in your movie, follow these steps:

1. Click the Add Files icon to open the Open dialog box, shown next. The Add Files icon is the open folder just to the right of the drop-down Library list box at the top of the library area.

2. Navigate to the folder that contains the files that you wish to import.

3. Select the files that you wish to import. Remember that you can use the CTRL and SHIFT keys to extend your selections in the standard Windows fashion.

4. Click the Open button to add the selected files to the library and to close the dialog box.

You can use the imported files just as if you had captured them in Roxio VideoWave Power Edition. In fact, once the files are in the Roxio VideoWave Power Edition library, there is no difference between imported and captured video files.

Using Scene Detection

You will probably find that most of your video clips would benefit from being broken down into scenes. When a clip has been broken down into scenes, you have the option of placing individual scenes on the storyline rather than using the entire clip.

As you will see shortly in the section "Using the Cutting Room," you can manually break clips into scenes. Often, this is the only way to get exactly the footage that you want for your movie. Still, it can be very handy to have Roxio VideoWave Power Edition automatically cut the clip into scenes saving you a lot of manual work.

Roxio VideoWave Power Edition has a special scene detection function that is used to automate the process. Figure 15-4 shows the Scene Detector dialog box that you use to break a clip into scenes.

To use scene detection, follow these steps:

1. Select the video clip in the library that you want to break into scenes.

2. Click the Scene Detector button on the Roxio VideoWave Power Edition toolbar along the left side of the window (this is the third button from the bottom). This will display the Scene Detector dialog box shown in Figure 15-4.

3. To use automatic scene detection, make certain that the Auto Detect option button is selected.

4. Click the Auto Detect Scenes button to begin the scene detection process. As scenes are detected, their thumbnails will appear in the box in the lower section of the dialog box. Scene detection happens as the video clip is played at a faster than normal speed, so you won't have to wait quite so long for the clip to play.

5. Click the Stop button when you want to finish the scene detection.

6. If the Scene Detector is too sensitive you will need to drag the Sensitivity slider to the left so that fewer scenes are detected. You can drag the slider to the right to increase the number of scenes that are detected.

7. If you are satisfied with the results, click the OK button to close the dialog box. Otherwise, delete the detected scenes, adjust the Sensitivity slider, and return to step 4 to redetect the scenes. You may have to do this several times to get the results that you really want.

You can also use the Scene Detector manually to split a clip into scenes by selecting the Manual option button. This method is less convenient than the

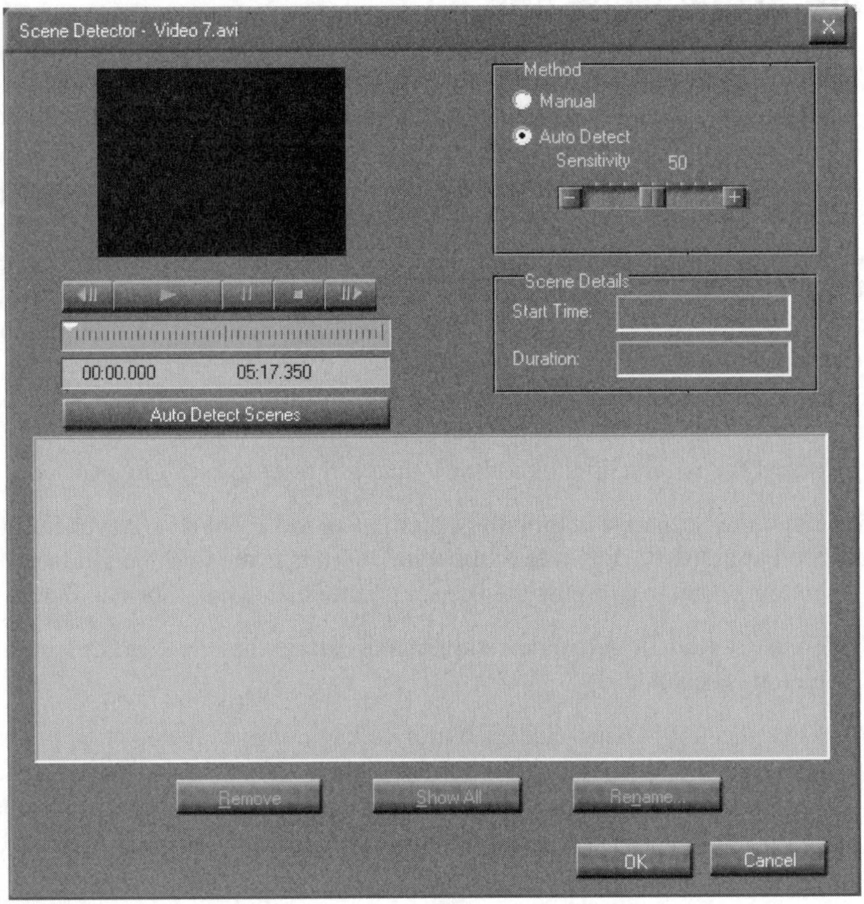

FIGURE 15-4 Use the Scene Detector to automate the process of breaking a clip into scenes.

automatic one, but it is generally faster than using the cutting room option for creating a large number of scenes within a single video clip.

> **TIP** *Be sure to click the Save button on the Roxio VideoWave Power Edition toolbar often to prevent loss of your work in case any problems should occur.*

Creating Your Movie

Once you have your video clips captured or imported and the scene detection process completed, you are ready to begin creating your movie. As with the other digital video editors we have seen earlier, this is largely a drag-and-drop process. Roxio VideoWave Power Edition does do some things a bit differently than the other editors, however, so you'll want to follow along in the next sections to learn how to make the best use of the program's features.

Adding Video to the Storyline

You begin the process of assembling your movie by selecting video clips from the library. You then drag-and-drop the clips onto the storyline (this is the Roxio VideoWave Power Edition term for *storyboard*). You can drop clips at the end of the storyline, or you can drop them between two existing clips if you want the new scene to appear in your movie between them.

One major difference you may notice while adding clips to the Roxio VideoWave Power Edition storyline is that you cannot drag clips to different locations once they have been added to the storyline. Therefore, rearranging the clips in your movie is a bit more complicated than you might expect—it pays to plan ahead and drop the clips onto the storyline in the correct order the first time around.

You may also be surprised to learn that you cannot simply drag a clip off of the storyline if you decide not to use that clip in its current location. So just how can you rearrange the storyline once clips have been added? The answer is that you must right-click the clip to display the pop-up menu shown here. Then you can use the editing commands on this menu to make adjustments. You can, for example, select Remove to take the selected clip completely off the storyline.

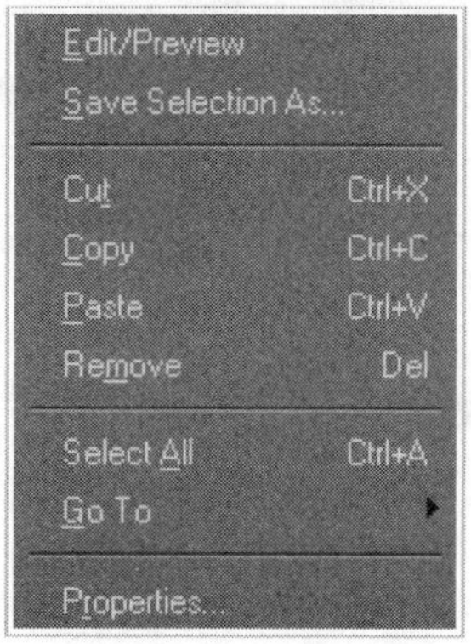

TIP *The storyline provides little information about the clips you have added, but you can quickly determine the length of any clip by double-clicking the clip to make it appear in the viewer window. The duration indicator below the playback controls will tell you the exact length of the clip.*

Incidentally, if you want to add specific scenes from a video clip to your movie rather than adding the entire clip, right-click the video clip in the library. Then select Show Scenes from the pop-up menu to display the scenes, as shown here. You can then drag-and-drop any of the scenes onto the storyline just as you would with an entire clip.

Editing Video Clips

Even though I am certain that your video clips are always perfect just the way they came from your camcorder, you will still probably want to know how to use the Roxio VideoWave Power Edition tools to make some adjustments to those clips. After all, even experts sometimes like to see if they can improve things by tweaking a little here and there.

You edit clips using the various tools that are displayed on the Roxio VideoWave Power Edition toolbar. It is possible to do your editing with clips before you add them to the storyline by selecting the clips in the library. If you want to edit a clip that you have added to the storyline, you will need to double-click the clip in the storyline in order to display that clip in the viewer window before you can edit the clip. It is really up to you which method you use. The primary difference between the two methods is that an edit of a clip within the library will appear in any copy of the clip that you drag-and-drop on the storyline, whereas editing a clip that is already on the storyline will affect only that single clip.

Using the Cutting Room

You use the Cutting Room tool to set the frames at which a clip begins or ends playing. You can also use this tool to extract a single frame as a bitmap image, to strip out the audio track, or (in some cases) to strip out the video track. The Cutting Room tool also enables you to select the frame that will be used as the thumbnail and to split a clip into two or more pieces.

To use the Cutting Room tool, click the Cutting Room button on the Roxio VideoWave Power Edition toolbar. When you do, the playback controls below the viewer window will add controls so you can mark the in-point and the out-point, as shown here (these are the controls that sit below the scrubber). The *in-point* is the frame where the clip begins playing, and the *out-point* is the frame where it stops playing.

To extract a part of the clip, set the thumbnail, or split the clip at the current frame, you use the controls at the lower-right side of the Roxio VideoWave Power Edition window, as shown here. Click the Apply button when you have made your changes, and the Close button when you are finished using the Cutting Room tool.

TIP	*If you want to make several splits in a scene, you will probably find that using the Scene Detector mentioned earlier in this chapter is somewhat easier to use than the Cutting Room tool.*

Using the Darkroom

The Darkroom tool enables you to adjust the visual appearance of a video clip. With this tool you can select from several different presets, or you can use sliders to control the brightness, contrast, color saturation, red balance, green balance, and blue balance.

To use the Darkroom tool, first select the clip that you want to adjust and then click the Darkroom button on the Roxio VideoWave Power Edition toolbar. You can then use the tools shown here (they appear in the lower-right corner of the Roxio VideoWave Power Edition window) to make manual adjustments:

You can also choose one of the presets from the library, as shown here. These preset combinations of adjustments are designed to quickly modify the appearance of the video clip in a specified manner.

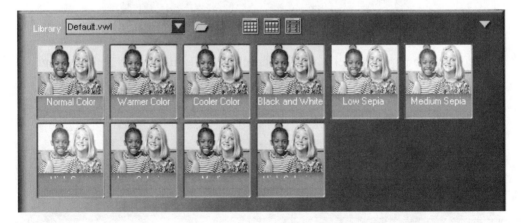

It is important to remember that any adjustments you make to the clip will apply to all of the frames within the clip. If you increase the brightness setting to compensate for an overly dark section of the clip, the whole clip will be set to the brighter setting. If this is not what you want, you may want to consider splitting the clip into two or more scenes so that you can apply your adjustments more selectively.

It is also pretty easy to get carried away with the manual slider adjustments and end up with a really bizarre effect. Small adjustments are usually the best. Here are just a few ideas about ways you might use the controls in the Darkroom:

- Increase the Blue level slightly if you want to give your movie a cold feel.

- Increase the Red level a little to make a scene seem warmer.

- Increase the Color level to compensate for the dull appearance of scenes that were shot on overcast days.

- Decrease the Brightness and increase the Contrast settings to give a scene the feel of a dark night.

- Increase the Contrast considerably to give a scene a cartoony, posterized effect.

When you have finished making your adjustments, click the Apply button and then the Close button to close the Darkroom.

Adding Special Effects

Roxio VideoWave Power Edition also offers a whole series of special effects that you can add to your movies. It is a little hard to describe all of the options you have available, but in Figure 15-5 you can see the special effects you can choose when you click the Special Effects button in the Roxio VideoWave Power Edition toolbar.

You may be somewhat surprised at just how much flexibility you have in applying special effects to your movies. For each of the special effects, you can choose from the following options:

- The starting frame where the special effect will begin to be applied to the selected scene (if you don't want it to begin at the first frame of the scene).

- The effective percentage level of the special effect when it starts. This controls the blending of the special effect and the normal frame at that point in the scene.

- A frame where the special effect begins holding at a specific level.

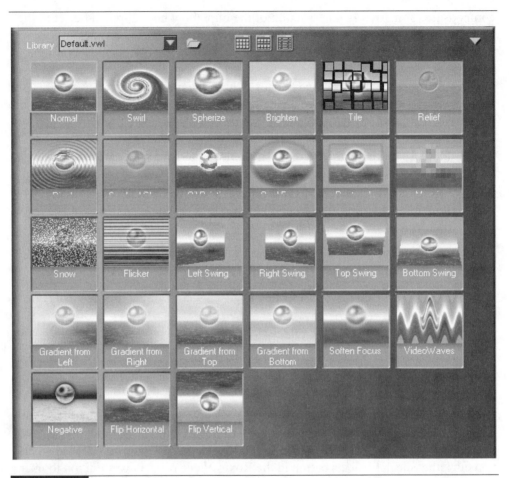

FIGURE 15-5 Choose the special effect that you want to use.

- The percentage level of the special effect during the holding period.
- A frame where the special effect stops the holding period.
- The frame where the special effect ends.
- The percentage level of the special effect at the frame where it ends.

These settings enable you to specify a gradual ramping up (or dropping off) of the special effect, a length of time when the special effect remains at a specific level, and then a gradual change until the special effect ends. To better understand how this might work, consider how the following scenario would appear:

■ You apply the Negative special effect to a scene, and allow it to begin with a level of 0 percent at the first frame.

■ One-third of the way through the scene you set the special effect to 100 percent and allow it to remain that way for the next third of the scene.

■ At two-thirds of the way through the scene you set the special effect to begin a decline back to 0 percent at the end of the scene.

Under this scenario, the scene would begin normally, but would gradually turn into a negative image of itself. Once it became fully negative, the scene would remain that way for a period before slowly returning to normal. In case you are still having trouble visualizing this, here I have captured a frame from a scene where I have applied the Negative special effect at 100 percent to the left side of the viewer window screen, while the right side of the screen appears normally. As you can see, applying the special effect changes the appearance of the scene considerably:

To apply a special effect to a scene, follow these steps:

1. Select the scene that you want to modify with a special effect.

2. Click the Special Effects button in the Roxio VideoWave Power Edition toolbar.

3. Double-click the special effect that you want to apply in the list of special effects area of the library.

4. If you want the special effect to delay its start until some point after the beginning of the scene, move the scrubber to that point and click the Set Effect Start Position button (the left-most button at the bottom of the playback controls area).

5. Drag the Effect Level slider to set the percentage level for the beginning of the special effect.

6. Move the scrubber to the frame where you want the special effect to begin holding at a constant level and then click the Set beginning of Effect Hold position button (the second button at the bottom of the playback controls area).

7. Click the Hold tab at the bottom of the Roxio VideoWave Power Edition window.

8. Drag the Effect Level slider to set the percentage level for the holding portion of the special effect.

9. Move the scrubber to the frame where you want the special effect to stop holding at a constant level and then click the Set end of Effect Hold position button (the third button at the bottom of the playback controls area).

10. Move the scrubber to the frame where you want the special effect to end and then click the Set Effect End Position button (the right-most button at the bottom of the playback controls area).

11. Click the Finish tab at the bottom of the Roxio VideoWave Power Edition window.

12. Drag the Effect Level slider to set the percentage level for the end of the special effect.

13. Click the Apply button.

14. Click the Close button to close the Special Effects tool.

You obviously won't be using special effects a lot in your movies, but they can be very useful in some situations. For example, the Mosaic or the Soften Focus special effect could be handy if you want to protect someone's identify in your movie, and the Tile special effect could be used to make it appear as though a scene were being assembled before your eyes. I'm sure that you will find many other uses for special effects.

Using TimeWarp

Another very interesting and fun feature that you will find in Roxio VideoWave
Power Edition is TimeWarp, a fast- or slow-motion effect. With this tool you can
make things appear to happen over a much shorter or a much longer period of time
than they really did. For example, a video clip that you shot while driving up a
mountain road might be sped up to give viewers a very dizzying experience, or you
might slow down a clip of someone running to make it appear as though they were
running into a very strong headwind.

   To use the TimeWarp tool, you click the TimeWarp button on the Roxio
VideoWave Power Edition toolbar. This will display the TimeWarp dialog box
shown here:

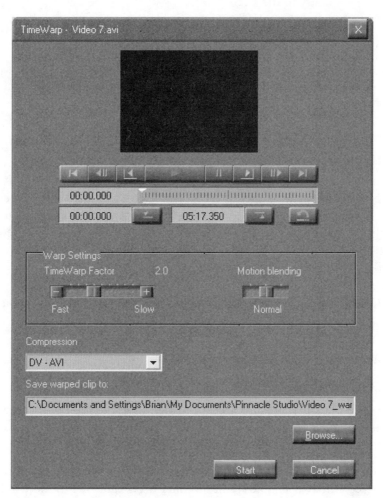

You will probably find most of the options in the TimeWarp dialog box to be quite simple to understand. Therefore, I'll just mention a couple of important items that you might otherwise overlook:

- You can use the Mark In Point and Mark Out Point buttons (just below the scrubber) if you want to apply the TimeWarp function to only a portion of the clip.

- The Motion blending setting smoothes the appearance of slow-motion effects, but it isn't needed for clips that are played faster than normal speed.

- It is best to use the DV-AVI compression, since this produces the best ratio of video quality to file size.

- It can take considerable time to produce the modified video clip, especially if you are applying the TimeWarp tool to a very long clip.

- Roxio VideoWave Power Edition stores both the original and the modified video clips in the library. You can easily identify the modified one because it has "warped" added to the name of the clip.

Adding Transitions

As you learned in Chapter 5, transitions can often help tell your story by making the change from one scene to another a bit clearer. Roxio VideoWave Power Edition offers you a very large number of transition effects that you can use in your movies. You are almost certain to find a transition that will fit your movie. Figure 15-6 shows the tools that you use to add transitions in Roxio VideoWave Power Edition.

To add a transition to your movie, follow these steps:

1. Click one of the transition areas between scenes on the storyline to select it. You can access the transitions only if you first select one of the transition areas.

2. Click the Transitions button on the Roxio VideoWave Power Edition toolbar to open the transitions tools, as shown in Figure 15-6.

3. Double-click the transition that you want to use in the library area to make it appear in the box in the middle of the lower area of the Roxio VideoWave Power Edition window.

4. If necessary, adjust the duration of the transition using the duration control between the two sample frames in the lower-right side of the window.

5. Use the playback controls to preview the transition.

6. Click the Apply button and then the Close button to close the transition tools.

TIP   *If the transition duration is too short, the first scene may turn to gray part way through the transition. Simply increase the duration slightly to create a smoother transition effect.*

Click here to open the transitions.

Add transitions between the clips on the storyline.

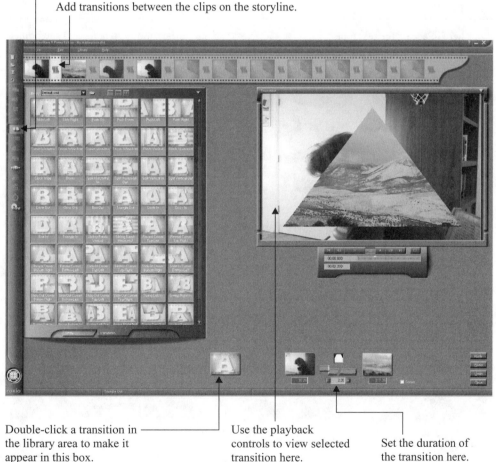

Double-click a transition in — the library area to make it appear in this box.

Use the playback controls to view selected transition here.

Set the duration of the transition here.

FIGURE 15-6   Here I've added a Triangle Out transition between the first two scenes.

Adding Text

Roxio VideoWave Power Edition is also quite adept at handling any text that you might want to add to your movies. In fact, Roxio VideoWave Power Edition offers some text animation options that you simply won't find in most digital video editors. You can, for example, create titles that roll up from the bottom of the screen, pause for a bit, and then scroll off to one side. In addition, the text can change color and fade in or out, if you like.

Figure 15-7 shows the tools that you use to add text to your movies. As with the special effects, any titles that you add can have a starting point, a holding

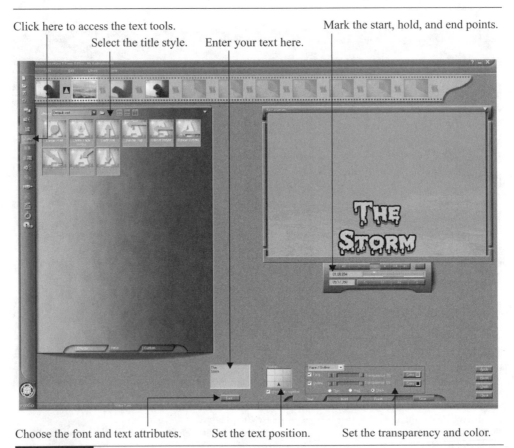

Click here to access the text tools.   Mark the start, hold, and end points.

Select the title style.   Enter your text here.

Choose the font and text attributes.   Set the text position.   Set the transparency and color.

FIGURE 15-7 These tools enable you to add titles to your movies.

section where the text does not move, and an ending section. You may want to refer back to the section on special effects if you need a bit of review about how those features work.

You should have little trouble adding titles to your movies at this point. The one new item that may confuse you slightly at first is the Position box. This box works along with the Start, Hold, and Finish tabs to control the motion of the titles. Each of the nine squares in the Position box shows the motion of the title at the start of the current section. In Figure 15-7, there is an arrow pointing into the box from the bottom center, and this indicates that the title will scroll up from the bottom at the start. The center square always holds an hourglass to indicate that the title will hold in the center of the screen (you would use this most often in conjunction with the Hold tab).

Adding Sound

As you are already aware, the video files that you capture from your camcorder generally include an audio track that is synchronized with the video. But even so, there are many times when you might want to add some additional audio components, such as background music or sound effects. To do so in Roxio VideoWave Power Edition, you use the Audio Studio tools.

In most ways, adding extra sound tracks to your movies is very similar to adding special effects or titles. This shows the lower-right section of the Roxio VideoWave Power Edition window as it appears when the Audio Studio tools are open:

You can use the Add Files button to add audio files to the library (on the Audio tab). If you add MP3 files, Roxio VideoWave Power Edition will automatically convert them into Wave files for use in your movies. You drag-and-drop the audio files into the Audio Tracks box in the Audio Studio tools section of the Roxio VideoWave Power Edition window. Once you have your audio tracks in place you have the following options (these options apply to the currently selected track in the Audio Tracks box):

■   Select Fade In or Fade Out to make the track slowly increase or decrease in volume.

■   Select Repeat to make the audio track loop as the video track plays.

- Select Mix to include the track or deselect it to silence the audio track.

- Use the Volume slider to set the volume level for the track.

- For tracks other than the recorded audio track that is associated with the video clip, click the Clip button to set the in and out points for the audio track.

Be sure to click the Apply button before you click the Close button—otherwise your selections will be discarded.

Burning Your Disc

When you have finished assembling your movie you can move on to creating your disc. This process, which involves adding menus as well as actually burning the disc, uses a different application than the main Roxio VideoWave Power Edition program that you have been using so far. You probably won't even notice the change since the two applications look so much alike and are so closely linked. Before you move on to that step, however, you have one more task to complete within the main Roxio VideoWave Power Edition application—producing your movie.

Producing Your Movie

In order to prepare your finished movie for burning to disc you must first select the File/Produce command from the main Roxio VideoWave Power Edition menu. This will display the Produce Movie dialog box, shown here, where you can select the output settings for rendering your movie:

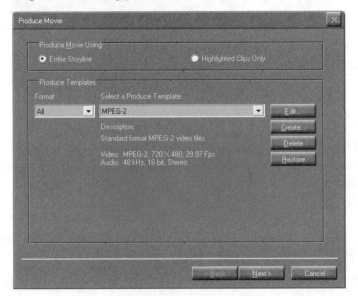

In most cases you will want to select the Entire Storyline option so that your entire movie is included in the output file. Then you must select the format and the template that will be used. You should select MPEG-2 if you are producing a DVD or S-VCD, and MPEG-1 if you are producing a VCD. For more information about some of the other output options, you may want to refer back to Chapter 9. Once you have made your selections, click the Next button to display the Summary Page dialog box, where you can specify a file name for your output file. After you have entered a file name, click the Produce button to begin the rendering process. At this point, you may want to go and have a cup of coffee since rendering takes considerable time (how much time depends on the complexity of your movie and the performance of your system).

Choosing the Output Format

When the rendering process is complete, you can move on to the final stages of creating your movie. Click the Author DVD button on the Roxio VideoWave Power Edition toolbar to display the New project settings dialog box shown here:

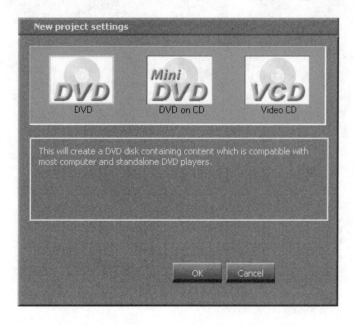

Click the output format that you want to use and then click OK. (Note that the Mini DVD or DVD on CD selection is the format that we have been calling S-VCD throughout the book.)

Adding Menus

Now you can begin adding your menus. Once again you will probably find that this process is quite simple, especially if you have read the chapters in Part II of this book. We will move through this topic fairly quickly so that you can get on with burning your movie to a disc without a lot of extra fuss. Figure 15-8 shows the tools that you will use to create your menus.

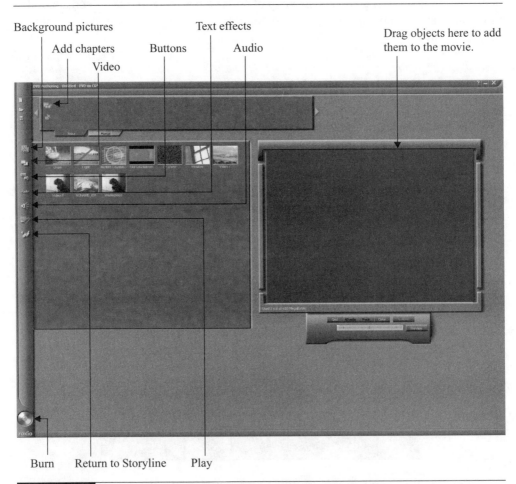

Background pictures   Text effects   Drag objects here to add
        Add chapters   Buttons   Audio   them to the movie.
              Video

Burn   Return to Storyline   Play

FIGURE 15-8   These are the DVD authoring tools.

Choosing Backgrounds

Begin your menu by clicking the Background pictures button. This will display a series of images that you can use as a background image. Choose the one that you want and drag-and-drop it into the viewer window.

> **TIP**   *If you want to use a solid color background rather than one of the background images, click the Color button below the viewer window and then choose your color. If you have already added a background image, you can remove it by right-clicking in the viewer window and choosing Delete Background from the pop-up menu.*

Adding Menu Buttons

Next, you need to add menu buttons to make it possible to select the video clips in your movie. You have two options here. If you want to create buttons that show a thumbnail image from the video clip, you can simply drag-and-drop a video clip into the viewer window. If you would rather have some other style of buttons in your menus, you first add the video clips and then drag-and-drop buttons onto those video clips in the menu. Here I have added three video clips to the menu, and I have dropped buttons onto two of them:

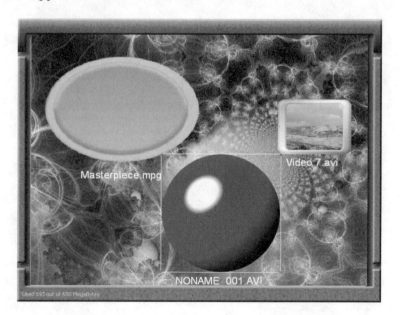

Click the Video button to select your video clips to add to the menu. If you want to replace the thumbnails with buttons, click the Buttons button to select from the button shapes.

In addition to simply creating menu buttons for each of the video clips, you may want to add chapters to some of the clips. This will enable someone who is viewing your movie to navigate using the chapter buttons on their remote control. To add chapter points, select the video clip that you want from the list of clips in the Titles area near the top of the Roxio VideoWave Power Edition window. Then click the Add chapters button to display the Edit chapters dialog box, as shown in Figure 15-9.

FIGURE 15-9 Add your chapter points using this dialog box.

Play the video clip (or drag the scrubber) until you reach a point where you want to add a chapter point. Click the Add button to add the chapter point. Continue this process until you have added all of the chapter points that you want. Then click the OK button to close the dialog box.

Adding Menu Text

You can add menu text (such as the name of your movie) by clicking the Text effects button to view the available text styles. Once again, you simply drag-and-drop your choice onto the menu to create a text entry.

You will likely want to change the default name that appears below the menu buttons, and you will certainly want to change the menu text from "abc" into something meaningful. Simply double-click any of the text in the menu to select it and you can then enter your new text to replace the existing text.

Adding Menu Sounds

If you want to have music playing while the menus are displayed, click the Audio button to display the list of audio files. Drag-and-drop the audio file that you want onto the menu in the viewer window.

TIP	*Right-click in the Audio pane and select Add Files to add your own audio files to the options shown.*

Finalizing the Disc

When you have finished creating your menus, you can finally move on to burning your disc. To do so, click the Burn button to display the Burn options dialog box shown in Figure 15-10. Choose the options that suit your needs. The first few times you create a disc you may want to select the Test and Burn option to make certain that the burn will be successful.

TIP	*If your recordable drive supports* burn-proof *technology, which greatly reduces the chance of errors in the burning process, make certain that you select the Burn-proof checkbox. You may need to check the documentation for your drive to learn if this is supported.*

Make certain that you have inserted the proper type of disc in the drive. Then click the Burn button in the Burn options dialog box to begin the burning process.

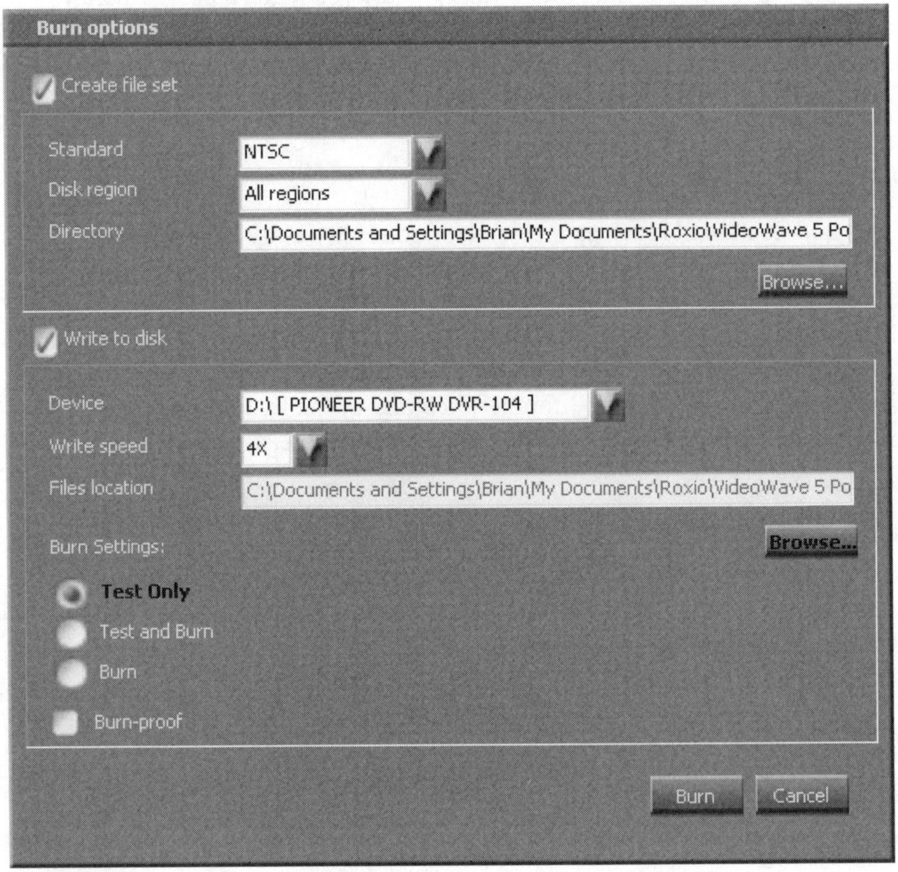

FIGURE 15-10 Choose your burn options and then create your disc.

We have now come to the end of our coverage of digital video editing software. In this chapter you learned how to create a complete movie in Roxio VideoWave Power Edition. Now it is time for you to go and have some fun creating your own DVD movies.

APPENDIX A

A New Type of Recordable DVD Drive

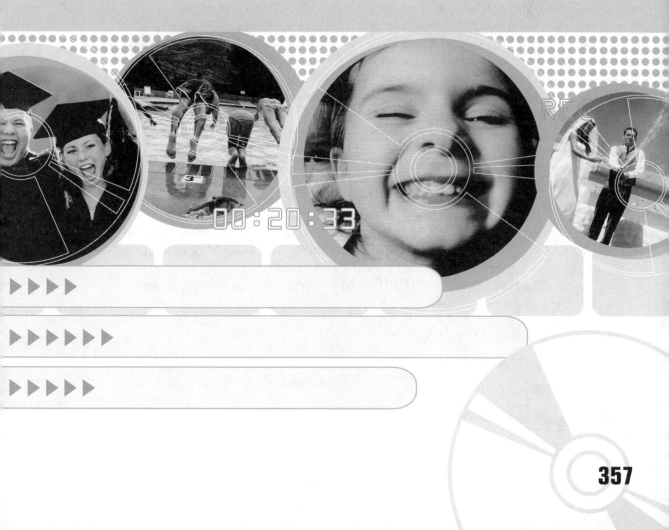

As I was in the process of writing this book, Panasonic released a new type of recordable DVD drive that I think you'll find quite interesting. It is the *LF-D521U DVD Multi* drive shown here:

As you would expect, the Panasonic DVD Multi drive can record on DVD-R, DVD-RW, CD-R, and CD-RW discs. It offers excellent performance on all of them, but so do a lot of other DVD-R/RW drives. What sets the Panasonic Multi drive apart is that this drive can also use DVD-RAM discs.

DVD-RAM discs are unlike any of the discs you currently use. In operation they function virtually like a hard drive. In effect, a DVD-RAM disc you buy for a few dollars becomes a 4.7 GB removable hard disk (or a 9.4 GB one if you buy double-sided DVD-RAM discs). What this means is that you can capture and store all of your raw video onto DVD-RAM discs. You can then edit your movies while they're still on the DVD-RAM disc, and when you're done creating your movie, you can burn it to an inexpensive DVD-R. Essentially, this means that you no longer have to worry about running out of hard disk space because you've filled it all up with your video content.

In case you're thinking that a DVD-RW disc seems to do just about the same thing as a DVD-RAM disc, think again. DVD-RAM discs act just like a hard disk, so you can add files, delete them, and write to the DVD-RAM disc in exactly the same way you now do to your hard drive. In fact, DVD-RAM discs are rated as capable of being written to at least 100,000 times!

The Panasonic LF-D521U DVD Multi drive has some pretty impressive performance specs too:

- 32X CD-ROM read speed

- 12X DVD-ROM read speed

- 12X CD-R write speed

- 8X CD-RW write speed

- 2X DVD-R write speed

- 2X DVD-RAM write speed

- 1X DVD-RW write speed

In addition, the drive comes with a full complement of very useful software, including the following:

- **Sonic MyDVD 3.5**   DVD authoring software

- **DVD-MovieAlbumSE 3**   Video recording, editing, and playback software

- **WinDVD 4**   DVD video playback software

- **B's Recorder GOLD5 BASIC**   Data and audio writing software

- **B's Clips**   Packet writing software (allows CD-RW and DVD-RW discs to be used like floppy disks)

- **FileSafe**   Automatic file backup software that enables you to schedule backups to your DVD-RAM discs

The Panasonic DVD Multi drive comes complete with a comprehensive manual, mounting screws, an internal audio cable, the software package just mentioned, a DVD-RAM disc, and a DVD-R disc. It is compatible with Windows 98, Windows Me, Windows 2000 Professional, Windows XP Home Edition, and Windows XP Professional.

If you need to add a recordable DVD drive to your system, or if you're thinking of upgrading your existing drive, the Panasonic DVD Multi drive should certainly be at the top of your list.

APPENDIX B

Using Your DVD Burner for Data and Music

How would you like to be able to get even more use from that DVD burner on your PC? You already know that your digital video editing software enables you to create your own DVD movies, but how about being able to do a whole bunch of more cool things? If the following list of possibilities sounds interesting, you may want to add a copy of *Roxio Easy CD & DVD Creator 6* to your system:

- Burn audio CDs and DVDs to create your own favorite mixes (with the potential for many hours of music on a single disc).

- Convert your audio CDs into MP3s so that you can have hundreds of hours of recordings on a disc.

- Create disc-based slide shows from all of your digital photos.

- Use recordable CD and DVD discs to store backups of your data files.

- Create bootable CDs and DVDs.

- Use drag-and-drop to copy files to recordable CDs and DVDs as easily as copying files to a hard drive or to a floppy disk.

- Create custom printed labels for your discs.

- Copy most CDs automatically.

Roxio Easy CD & DVD Creator 6 is the latest version of some of the best known disc burning software that is available for Windows. It includes all of the capabilities I mentioned plus quite a few more—it even includes a simple DVD movie creation application called *DVD Builder*. The following illustration shows the main menu that you can use to choose one of the applications that make up this suite of disc burning tools.

Have you ever thought about converting your old vinyl recordings into audio CDs? If you have ever tried this, you know that those old records were pretty noisy compared to modern recordings. Here I'm using one of the Roxio Easy CD & DVD Creator 6 tools (*Sound Editor,* which is a part of the *Audio Central* application) to remove the clicks and pops that so often mar those recordings:

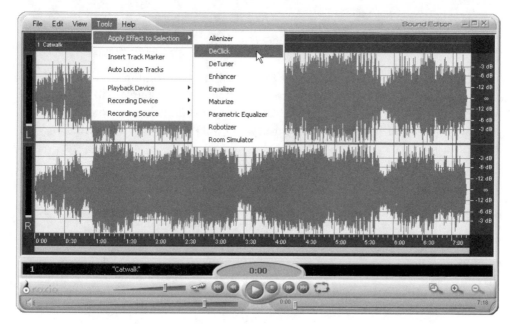

If you are at all like me, you probably have hundreds of photographs that you should be sharing with other family members or friends, but you have just never come up with a really good method of sharing them. If so, then you will find that Roxio Easy CD & DVD Creator 6 has a solution for that problem too. As this shows, *PhotoSuite 5* allows you to easily organize and share your photos in a number of different ways, including burning them to discs:

I could go on about all of the great features that you'll find in Roxio Easy CD & DVD Creator 6, but it may just be best to tell you to visit the Roxio Web site (www.roxio.com) for more information. No matter what digital video editing program you use, I think you'll find that Roxio Easy CD & DVD Creator 6 belongs on your PC too.

Index

INTERNATIONAL CONTACT INFORMATION

AUSTRALIA
McGraw-Hill Book Company Australia Pty. Ltd.
TEL +61-2-9900-1800
FAX +61-2-9878-8881
http://www.mcgraw-hill.com.au
books-it_sydney@mcgraw-hill.com

CANADA
McGraw-Hill Ryerson Ltd.
TEL +905-430-5000
FAX +905-430-5020
http://www.mcgraw-hill.ca

**GREECE, MIDDLE EAST, & AFRICA
(Excluding South Africa)**
McGraw-Hill Hellas
TEL +30-210-6560-990
TEL +30-210-6560-993
TEL +30-210-6560-994
FAX +30-210-6545-525

MEXICO (Also serving Latin America)
McGraw-Hill Interamericana Editores S.A. de C.V.
TEL +525-117-1583
FAX +525-117-1589
http://www.mcgraw-hill.com.mx
fernando_castellanos@mcgraw-hill.com

SINGAPORE (Serving Asia)
McGraw-Hill Book Company
TEL +65-6863-1580
FAX +65-6862-3354
http://www.mcgraw-hill.com.sg
mghasia@mcgraw-hill.com

SOUTH AFRICA
McGraw-Hill South Africa
TEL +27-11-622-7512
FAX +27-11-622-9045
robyn_swanepoel@mcgraw-hill.com

SPAIN
McGraw-Hill/Interamericana de España, S.A.U.
TEL +34-91-180-3000
FAX +34-91-372-8513
http://www.mcgraw-hill.es
professional@mcgraw-hill.es

**UNITED KINGDOM, NORTHERN,
EASTERN, & CENTRAL EUROPE**
McGraw-Hill Education Europe
TEL +44-1-628-502500
FAX +44-1-628-770224
http://www.mcgraw-hill.co.uk
computing_europe@mcgraw-hill.com

ALL OTHER INQUIRIES Contact:
McGraw-Hill/Osborne
TEL +1-510-596-6600
FAX +1-510-596-7600
http://www.osborne.com
omg_international@mcgraw-hill.com

save the day

▷📷 DVD VIDEO

Use DVD Builder™ to turn home movies into DVDs, complete with transitions, music and animated menus. Then share them today, tomorrow or twenty years from now.

📷 PICTURES

Digital pictures never fade. Use PhotoSuite® 5 to organize, edit and burn all your family photos to CD, so your precious memories will be safely preserved for the long haul.

🎵 MUSIC

Use AudioCentral™ to create your own personal CD music mixes, edit tracks, add effects and cross-fades or play, organize and burn original compositions.

🗁 DATA BACK-UP

Birth certificates, insurance records, tax information—archive all those large important family files to multiple discs and keep them organized and accessible.

INTRODUCING NEW EASY CD & DVD CREATOR™ 6 FROM ROXIO.®

Completely redesigned to be even more intuitive and capable, Easy CD & DVD Creator 6 is the most powerful and easy to use suite of digital media software out there. Use it to capture, manage, edit, and burn photos, video, music and data to CD or DVD. Then keep all your stuff organized at your fingertips for easy access whenever you need it with Easy CD & DVD Creator 6, from Roxio. To learn more check out roxio.com or visit your local software retailer.

About the CD

Included with this book is a free CD from Sonic Solutions. This trial disc contains the following product samples:

- **CinePlayer™ 1.5** Software DVD player allows you to view DVDs on your PC

- **MyDVD® Plus 4.0.4** Entry-level DVD-Video and VCD authoring

- **DVDit!® SE 2.5.5** Mid-level DVD-Video authoring

- **RecordNow™ MAX 4.56** CD and DVD burning

- **Backup MyPC™ 4.95** Data backup and recovery

- **ReelDVD® 3.0.3** Professional level DVD-Video authoring

We hope you enjoy using these products. Each application, except ReelDVD, will expire 30 days after installation. ReelDVD expires after 14 days. You will have an opportunity to purchase MyDVD, DVDit!, and ReelDVD through a Purchase button in the autorun installer. You can also purchase the latest version of any application by going to the Sonic estore (http://estore.sonic.com) and placing your order.

You will find more information about these applications, installation, and purchasing on the CD's ReadMe file.